Electronic Components and Technology

Engineering Applications

S.J. Sangwine
Department of Engineering
University of Reading

 Van Nostrand Reinhold (UK) Co. Ltd

First published in 1987 by
Van Nostrand Reinhold (UK) Co. Ltd
Molly Millars Lane, Wokingham, Berkshire, England

Typeset in Times 10 on 12pt by
Colset Private Ltd, Singapore

Printed and bound in Hong Kong

British Library Cataloguing in Publication Data

Sangwine, S.J.
    Electronic components and technology:
    engineering applications. — (Tutorial
    guides in electronic engineering; 13).
    1. Electronic apparatus and appliances
    I. Title   II. Series
    621.3815′1   TK7870

    ISBN 0–278–00017–7

ISSN 0266 2620

Contents

Preface

This book is intended to support Engineering Applications studies in electronic engineering and related subjects such as computer engineering and communications engineering at first and second year undergraduate level. *Engineering Applications*, abbreviated to EA, is a term first used in the report of the Finniston inquiry into the future of engineering in the United Kingdom. Finniston used the terms *EA1* and *EA2* to refer to the first and second elements of a four-stage training in Engineering Applications, to be taken as part of a first degree course in engineering. Later, the Engineering Council, established as a result of the Finniston report, expressed the concept of EA1 as

> An introduction to good engineering practice and the properties, behaviour, fabrication and use of relevant materials and components.

and EA2 as

> Application of scientific and engineering principles to the solution of practical problems of engineering systems and processes.

These quotations are taken from *Standards and routes to registration*, otherwise known as SARTOR, published by the Engineering Council in December 1984. They are reproduced here with the permission of the Engineering Council.

Although EA studies should be integrated into the fabric of a degree course, there is a need to draw out elements of practice to provide emphasis. It is intended that this book should be used as a source, complementing other texts, for such studies.

Part of the aim of this book is to inform the reader about components, technology and applications, but it is also intended to create an awareness of the problems of electronic engineering in practice. I hope the reader of this book will be encouraged to tackle these problems and go on to become a competent and professional electronics engineer.

In the context of electronics, product design is an activity that begins, by and large, with *components* rather than *materials*. This is not to say that a study of materials is not relevant as a part of EA1, but as there are many existing texts covering the subject, *materials* has been excluded from this book in favour of more coverage of components.

The book begins with an introduction to electronic interconnection technology including wiring, connectors, soldering and other jointing techniques, and printed circuits. Chapter 3 is devoted to the very important technology of integrated circuits, concentrating on their fabrication, packaging and handling. *Components* is taken to include power supplies, as in many applications a power supply unit is bought-in as a sub-system. The main characteristics of power supplies and batteries are covered in Chapter 4. Passive electronic components are introduced in Chapter 5, and with them the book begins to include a major theme developed over the following two chapters: the *parasitic effect*. This includes the non-ideal properties of passive components introduced in Chapter 5; heat and its management in Chapter 6 and parasitic electromagnetic effects in Chapter 7. Chapter 8 reviews good engineering practice in relation to reliability and maintainability, two important aspects of design which unfortunately, are often overlooked by electronic circuit designers. Chapter 9 introduces environmental influences on electronic products and the subject of testing both for environmental endurance and in production. The final chapter in the book introduces safety.

The book assumes that the reader has taken the first one or two terms of a degree course, although some of the earlier material could be studied sooner. Extensive cross-references to more specialized texts have been given in the marginal notes and the bibliography, including, where appropriate, references to other texts in the Tutorial Guides Series.

Part of the aim of this book is to inform the reader about components, technology and applications, but it is also intended to create an awareness of the problems of electronic engineering in practice. I hope the reader of this book will be encouraged to tackle these problems and go on to become a competent and professional electronics engineer.

Safety Note

The material in Chapter 10 is of course at an *introductory level only*, and readers are cautioned that professional competence in safe electrical design cannot be achieved merely by studying the contents of this chapter.

Acknowledgements

This book developed from a lecture course which I first presented in 1985 as a part of new EA material introduced into engineering courses at Reading. I would like to thank my colleague Peter Atkinson for his early suggestion that a book could be written and for his help and encouragement during the past two years. I would also like to thank Mr S.C. Dunn who has recently retired from the position of Chief Scientist at British Aerospace and who made many helpful suggestions at an early stage, including the theme of parasitic effects. The following people contributed advice, criticism or technical background and I wish to acknowledge their help: Susan Partridge of GEC Hurst Research Centre for reading the first draft of Chapter 3; Alistair Sharp of Eurotherm Ltd. for help with the case study in Chapter 2; John Barron of Tectonic Products, Wokingham for help with the subject of printed circuits; John Terry of the Health and Safety Executive and Ken Clark, Deputy Director at BASEEFA, both for help and criticism of Chapter 10; Dr George Bandurek at Mars Electronics at Winnersh for commenting on Chapter 8; Martin Thurlow of the Electromagnetic Engineering Group, British Aerospace and David Hunter of the Army Weapons Division, British Aerospace for assisting with illustrations and background. I am also grateful to Carole Hankins for typing the manuscript and to my wife Elizabeth Shirley for her support over many months of writing and revising. Professor A.P. Dorey was consultant editor for this book and made many helpful comments on drafts of the manuscript.

Several companies and organizations have supplied illustrations or granted permission for me to use their copyright material and they are acknowledged in the text. Extracts from British Standards are reproduced by permission of the British Standards Institution. Complete copies of British Standards can be obtained from BSI at Linford Wood, Milton Keynes, MK14 6LE, UK.

Introduction 1

Modern electronic engineering products are found in a wide range of applications environments from the floor of the deep ocean (submarine cable repeaters) to geostationary orbit (microwave transceivers on board communications satellites), from the factory floor (industrial process controllers and numerically controlled machine tools) to the office (word processors). They can be found in the home (audio and video systems and microwave ovens), in schools (microcomputers and pocket calculators), in hospitals (tomographic scanners), inside the human body (heart pacemakers) and under car bonnets (electronic ignition). These products may be mass-produced by the thousand or they may be one-off special systems. They may be intended to last for decades or they may be designed deliberately for a short life. They should all be fit for their intended purpose, and hopefully they will also yield some benefit to their users and their manufacturers.

All electronic products depend on the physical and electrical properties of insulating, conducting and especially semiconducting *materials*, but by and large, the designer of an electronic product works with *components* and *technologies*, such as integrated-circuit (IC) technology, rather than with basic materials. A critical aspect of product design is the interconnection of components and for this reason this book starts with a chapter covering the technology of interconnection. The technology of interconnecting electronic components, circuits and subsystems has, until now, often been neglected in electronic engineering texts at degree level. It is true that the detailed layout of a printed circuit board (PCB) is not a task likely to be undertaken by a graduate engineer. Nevertheless a PCB has electrical properties and its design, together with the choice of components to go on it, can have a significant effect on the performance, the cost of production, the production yield and the reliability and maintainability of the assembled board, and quite likely the product of which it is a part. Jointing techniques, especially soldering, are of tremendous importance in electronic engineering and solder is an engineering material which should be specified as carefully as a mechanical engineer specifies structural steel: what *type* of solder is best suited to a particular application? In many cases just 'solder' will not do.

The third chapter deals with IC technology. At present, few engineers are involved in high-volume IC design, but in the future an increasing number will design or use semi-custom ICs. Consequently, the treatment in this book is not for the IC specialist: it is aimed at the much larger group of electronics engineers who will be using ICs or designing a gate-array or standard-cell IC of their own.

The competent electronics engineer needs a good understanding of the components and sub-systems from which he will construct his designs. Not only must he be aware of their ideal behaviour, he must understand how they are fabricated in order to appreciate their performance limitations. The next two chapters, therefore, cover power sources and power supplies (an important class of electronic sub-system) and the subject of passive electronic components.

The study of electronic components introduces the third major theme of this book: the *parasitic* effect. Real electronic components and circuits, as opposed to *ideal* ones, possess parasitic properties which are incidental to their intended properties. A wire-wound resistor, for example, is also inductive and has an impedance which varies with frequency. *Heat* is produced in significant quantity in some electronic systems and positive design measures often have to be taken to remove it. Electromagnetic energy can radiate from electronic circuits and couple into other circuits, causing faulty operation. A chapter has been devoted to this and other parasitic electromagnetic effects. This book does not attempt to cover all possible parasitic effects: to do so would be impossible even in a much larger book and would serve little useful purpose. Electronics engineers must expect parasitic effects and take them into account when designing electronic products.

So far this introduction has dealt with matters that affect design and performance in ways that are important at the beginning of the life of a product. Without an understanding of components, technology and parasitic effects, the design engineer will not be able to design good products that meet the required level of performance at the required cost. Many electronic products, however, will have a life that lasts far longer than the designer's interest in the design. It is during the operating life of a product that long-term effects become important. Components and materials *age*: they deteriorate physically and chemically and ultimately they *fail*. The study of these problems and the prediction of product life is known as *reliability*. Not surprisingly, the reliability of a product can be influenced by its design, for better or for worse, and if a product is capable of being repaired, the ease and expense with which it can be restored to working order can also be affected by decisions taken at the design stage. Reliability can also be influenced by a product's operating environment. Did the designer consider the effect of temperature, humidity, corrosion and dust? Is there some unknown environmental factor which will doom his product to early failure? As with parasitic effects, after introducing some of the many environmental hazards to electronic equipment, this book leaves the reader to consider what the problems of his product's environment might be.

Lastly, this introduction has dealt with the electronic product itself: will it work and continue to work for long enough? Will it succumb to environmental stress? Engineers must also look at their designs from another viewpoint: will they do anyone any harm? All design engineers, including those working in electronic engineering, have a professional and statutory duty to consider safety when designing products. The final chapter in this book introduces the subject of safety in electronic engineering.

Interconnection Technology   2

□ To emphasize the importance of interconnection in electronic product design.

□ To discuss jointing technology, especially soldering and solderless wire-wrapping.

□ To outline the main types of discrete wiring and cabling.

□ To describe the technology of printed circuits.

□ To give an introduction to rework techniques.

□ To present a short case study illustrating the importance of interconnection in industrial product design.

Objectives

All except the smallest of electronic systems are built up from sub-systems or sub-assemblies which are in turn built from electronic components such as resistors, capacitors, transistors, integrated circuits, displays and switches. A minicomputer, for example, is built from a power supply sub-system, a central processor unit (CPU) and a number of peripheral sub-systems such as disk and tape drives. Small self-contained electronic products such as pocket calculators are often built directly from components with no identifiable sub-systems.

From the lowest level of component up to the system level, the constituent parts of an electronic system have to be interconnected electrically. The lowest level in the hierarchy of interconnection is the electrical *joint*. From the very earliest days of electronics, long before the invention of the transistor and integrated circuit, soldering has been an important technique for making electrical joints. Hand soldering is still widely used in prototype work, repair work and to a lesser extent in production. It is one of very few manual skills needed by engineers at all levels within electronic engineering. Not all electrical joints in an electronic system need to be soldered: the technology of solderless wire-wrapping is well established in digital electronics, both for prototype and production wiring, and joints can also be made by welding or crimping. Components and sub-systems are interconnected by wiring which can be in the form of either discrete wires and cables or printed circuits. A short case study at the end of this chapter illustrates the modern trend in electronic engineering towards printed circuit interconnection wherever possible, avoiding discrete wiring because of the high cost of assembling and inspecting individual wires.

The printed circuit board, or PCB, is tremendously important in almost all application areas of modern electronics. Not only does it provide a cheaply mass-produced means of interconnecting hundreds or thousands of individual components, it also provides a mechanical mounting for the components.

Compare the modern printed circuit board with the discrete wiring underneath a valve radio or television chassis and the work involved in bending, punching and drilling aluminium sheet for a valve chassis.

Jointing

Several techniques are used in electronic engineering for making electrical joints. The most important of these is soldering which is used mainly for attaching and jointing components to PCBs, but is also widely used for jointing in cable connectors. Another important technology is solderless wire-wrapping, which finds application in prototype wiring for logic circuits and in production wiring of backplanes interconnecting PCBs. Welding is used in some specialized electronic applications and is a very important jointing technique in integrated circuit manufacture. Finally, in applications where soldering or welding cannot be used, mechanical crimping can make sound electrical joints.

Soldering

Strictly, a solder need not be a tin–lead alloy, but when the term *solder* is used in electronics it nearly always refers to an alloy based mainly on tin and lead.

Solder is a low-melting-point alloy of tin and lead, used for making electrical and mechanical joints between metals. Soldering does not melt the surfaces of the metals being joined, but adheres by dissolving into the solid surface. Figure 2.1 shows an idealized cross-section of a soldered joint illustrating this point. Soldered joints can be made by hand using an electrically-heated soldering iron or by a mass-soldering process in which all the joints on a PCB are made in one automated operation. Both techniques are important and are described in detail below and in a later section of this chapter.

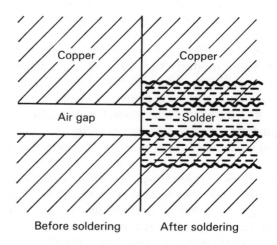

Fig. 2.1 Diagrammatic representation of a soldered joint (not to scale).

Commercial solders are available with several different proportions of tin to lead and with traces of other metals to enhance their properties. Tin and lead are soft metals with melting points of 232°C and 327°C respectively. Alloys of these metals generally start to melt at a temperature of 183°C, which is lower than the melting temperature of either pure metal. Figure 2.2 shows a simplified phase diagram for tin–lead alloys. The ratio (by weight) of tin to lead is plotted horizontally with 100% lead on the left and 100% tin on the right. The vertical scale represents temperature. In the top region of the diagram above the line extending from the melting point of lead at 327°C across to the melting point of tin at 232°C via the point at a tin lead ratio of 63:37 and a temperature of 183°C, the alloys are liquid. The bottom region of the diagram represents the ranges of temperature and tin : lead ratio over which the alloys are solid. The two triangular regions represent

Phase diagrams in general and the tin–lead phase diagram in particular are discussed by Anderson *et al.*

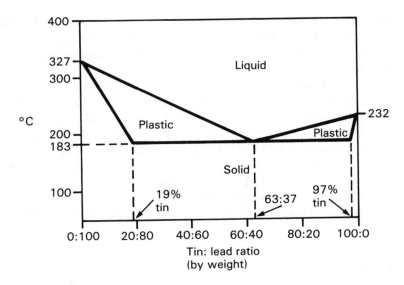

Fig. 2.2 Phase diagram for tin–lead solder alloys (simplified).

temperatures and compositions where the alloys are in a plastic state consisting partly of solid and partly of liquid. The alloy with a tin : lead ratio of 63:37 is the only one which changes sharply from solid to liquid at a single temperature. This alloy is known as a *eutectic* alloy. It is fully liquid at the lowest possible temperature for a tin–lead alloy.

For general electronic jointing, a 60:40 solder is used, which is fully molten at about 188°C. 40:60 solder, which is fully molten at about 234°C, is also readily available.

Four requirements must be met if a good soldered joint is to be made and an understanding of these is essential in order to develop skill at hand soldering. Firstly, the surfaces to be joined must be solderable. Not all metals are readily soldered without special techniques. Aluminium, for example, is an extremely reactive metal which rapidly oxidizes on exposure to air to form a passivating oxide layer which prevents solder from alloying with the underlying metal. Gold is very easily soldered because the metal does not oxidize. Copper and brass oxidize readily but the oxide layer can be removed easily, so that these metals are solderable. The second requirement for a good soldered joint is cleanliness: the surfaces to be joined must be free of grease, dust, corrosion products and excessively thick layers of oxide. In most electronics applications the surfaces to be soldered will have been plated with gold or tin–lead alloy during manufacture in order to provide a readily solderable surface. Coating a metal with solder is known as *tinning*, and protects the metal from oxidation. Cleaning is not usually needed, therefore, unless a component or PCB has been stored for a long time or become contaminated with grease or dirt. The third requirement is that any layer of oxide on the surfaces must be removed during soldering and prevented from regrowing until the molten solder has alloyed with or 'wetted' the surface. This is achieved by a *flux* which chemically removes the oxide layer and reduces surface tension, allowing molten solder to flow easily over the surfaces to be joined.

The proportion of tin (by weight) is conventionally stated first.

It is possible to demonstrate the reactive nature of aluminium by removing the oxide layer with a little mercury rubbed onto the aluminium surface. The freshly exposed surface reacts rapidly with moisture in the air and the metal becomes hot.

PCBs normally incorporate a solderability test pad in an unused corner of the board, so that the solderability of the board can be checked before component assembly.

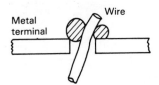

Globular solder with no 'wetting' of surfaces

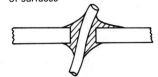

Correctly 'wetted' solder joint

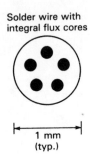

Solder wire with
integral flux cores

|← 1 mm →|
(typ.)

For hand soldering, solder wire with integral cores of flux is used. The fluxes commonly used for electronic work are rosin-based and chemically mild, leaving a non-corrosive residue. More powerful fluxes based on acids may be needed on less solderable metals, but must be thoroughly cleaned off afterwards because they leave a corrosive residue. The final requirement for a good soldered joint is heat. Both surfaces to be joined must be heated above the solidification temperature of the solder, otherwise the solder will chill on contact and will fail to flow evenly and alloy with the surfaces. When soldering by hand, heat transfer from the soldering iron to the joint is improved if there is a little molten solder on the tip of the soldering iron. The joint should be heated with the iron, and the solder wire applied to the joint (not the iron). The iron should not be removed until the solder has flowed into the joint. It is very important that the joint is not disturbed until the solder has fully solidified, otherwise a high resistance (dry) joint will result from mechanical discontinuities in the solder.

Wire-wrap Jointing

Soldering is a good technique for mass jointing on PCBs, but has several disadvantages for discrete wiring joints. An alternative jointing technique exists for logic circuits and low-frequency applications known as solderless wire-wrap. Special wire and terminal pins are used for wire-wrap jointing and special hand-operated or electrically powered tools are required. Various types of IC sockets and connectors, including PCB edge connectors, are made with wire-wrap terminal pins. Wire-wrap interconnection is used for prototype and production wiring of logic boards and for production wiring of backplanes for interconnecting a rack of PCB sub-units. It is a faster technique than soldering, requires no heat and

Wire-wrapping was first developed at Bell Telephone Laboratories in the USA about 1950.

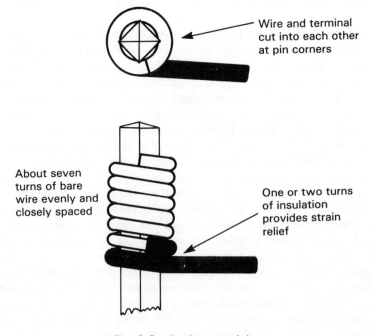

Wire and terminal cut into each other at pin corners

About seven turns of bare wire evenly and closely spaced

One or two turns of insulation provides strain relief

Fig. 2.3   A wire-wrap joint.

produces no fumes, and can easily interconnect terminals spaced as little as 2.5 mm apart. Figure 2.3 illustrates a typical joint. The terminal pin is typically 0.6–0.7 mm square and 15–20 mm long. The wire is solid and about 0.25 mm in diameter. The joint consists of about seven turns of bare wire and one to two turns of insulated wire, wound tightly around the terminal pin. At each corner of the terminal pin, the wire and the pin cut into each other, making a metal-to-metal connection which improves with age through diffusion of the two metals. The wire is under tension and slightly twists the square pin. The insulated turns of wire act as a strain relief at what would otherwise be a weak point of the joint.

The tools required for making wire-wrap joints are more expensive than those needed for making soldered joints, but the time saved in production work soon covers the cost of the tools. The wire is stripped using a special tool which ensures that the correct length of insulation is removed. The stripped end of the wire is then inserted into the offset hole of a wrapping bit which may be part of a hand-operated or powered tool. The centre hole of the wrapping bit is then slipped over the terminal pin and the tool rotated evenly to wrap the wire around the pin. Most wire-wrap terminal pins have enough length for up to three joints.

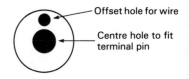

Offset hole for wire

Centre hole to fit terminal pin

End view of a wire-wrap bit.

One of the very significant advantages of wire-wrapping for prototype wiring is the ease and speed with which joints can be unwrapped, to allow circuit modifications. A special tool is required to unpick a joint, and any joints further out along the pin have to be unpicked first. For this reason, when a group of pins are to be connected together each wire should be at the same level on both pins connected. Figure 2.4 shows the right and wrong ways of connecting together a series of pins. In the 'daisy-chained' arrangement a change to one wire can often require many other wires to be unpicked and remade. Because of the cutting action of the wire on the corners of the terminal pins, there is a limit to the number of times that a pin can be rewired. The small number of modifications likely to be needed to a prototype are not likely to cause problems with poor joints, but continual reuse of wire-wrap IC sockets could be more trouble than the sockets are worth. The wire removed from an unpicked joint is not reusable, so that in nearly all cases, both ends of a wire have to be unpicked. The unpicked wire should be

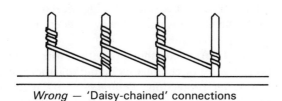

Wrong — 'Daisy-chained' connections

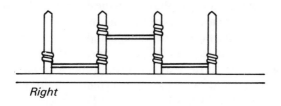

Right

Fig. 2.4   Right and wrong ways of wire-wrapping a group of terminals.

carefully removed from the circuit and thrown away, taking care that fragments of stripped wire do not fall back into the circuit. An intermittent short circuit caused by a piece of loose wire can take a long time to diagnose.

One final point about wire-wrapped connections is that there is little point in trying to arrange the wires into tidy bundles: direct point-to-point wiring can be easier to inspect and check and minimizes problems with crosstalk.

Welding

Welding is a jointing technique where two metal surfaces are placed in intimate contact and then fused together by melting both surfaces. Heat can be applied by a flame, thermal conduction or electric heating. There are limited applications of welding in electronics, except in the bonding of ICs to their packages. External connections from bonding pads on an IC are made by attaching fine gold wires using either ultrasonic welding or thermocompression bonding as described in the next chapter.

Crimping

A fourth jointing technique used in electronics (and much more extensively in electrical engineering) is crimping. A crimped connection is made by crushing a special terminal onto a wire of the correct size using a purpose-made tool. The wire is gripped mechanically by the crushed terminal. Electrical contact depends on the mechanical integrity of the joint. Unlike a wire-wrapped or soldered joint, the electrical connection is not gas-tight and can therefore be prone to corrosion. Crimping is an especially useful technique for jointing unsolderable wires, and for rapid wiring assembly in production.

Crimping is very widely used in the automotive industry for rapid assembly of vehicle wiring harnesses.

Discrete Wiring

Modern electronic product designers tend to avoid using discrete wiring in favour of printed-circuit interconnection because of the high cost of hand jointing and the likelihood of errors. Some wiring is nearly always needed, however, especially in larger systems.

A *wire* is a single or multi-stranded conductor with or without insulation whereas a *cable* is a collection of wires or conductors bound together, possibly with some overall insulation or other protection. *Equipment wire* used in electronics is normally of copper or tinned copper with either PVC or PTFE insulation. PVC-covered wire can be used at up to 70°C. Above this temperature PTFE-covered wire must be used. The size of a wire can be stated either as a cross-sectional area (in mm^2) or as the number of strands followed by the diameter of each strand (in mm). Thus 7/0.2 represents seven strands of 0.2 mm diameter.

You may still find wire sizes quoted in British Standard Wire Gauge (SWG) or in American Wire Gauge (AWG). You will need a table of wire gauges to find out the cross-sectional area.

For many electronic purposes, the voltage and current ratings of the lightest equipment wire are far greater than the requirements of the circuit. A 7/0.2 PVC-covered wire, for example, is rated at 1.4 A (this is fairly small wire). This rating is conservative (the wire could carry a greater current without overheating) to allow for the possibility of several wires being bundled together in a confined space. A 7/0.2 PTFE-covered wire is rated at 6 A because the insulation can withstand

Table 2.1   Current ratings of copper equipment wire

(a) PVC insulated

Construction	Cross-sectional area (mm²)	Current rating (A)
7/0.2	0.22	1.4
16/0.2	0.5	3.0
24/0.2	0.75	4.5
32/0.2	1.0	6.0
7/0.53	1.5	17.0

(b) PTFE insulated

Construction	Cross-sectional area (mm²)	Current rating (A)
7/0.15	0.12	3.5
7/0.2	0.22	6.0
19/0.16	0.38	9.0

higher temperatures caused by self-heating. As the following worked example shows, however, there will be a considerable voltage drop along wire of this cross-section carrying 6 A.

Calculate the voltage drop along 500 mm of 7/0.2 PTFE-insulated copper equipment wire carrying a current of 6 A. The resistivity of copper is 1.7×10^{-8} Ωm.   **Worked Example 2.1**

Solution   The cross-sectional area, A, of the wire is $7\pi \times (0.1)^2$ mm² or 0.22 mm². The resistance of a length, l, resistivity, ρ is

$$R = \frac{\rho l}{A}$$

Hence the resistance of the 500 mm length is

$$1.7 \times 10^{-8} \times (0.5)/0.22 \times 10^{-6} \cong 40 \text{ m}\Omega$$

From Ohm's law, the voltage drop is about 240 mV. If this wire is used to connect a power supply to a load only half a metre away, the voltage at the load will be 0.48 V less than the voltage at the terminals of the power supply (there will be a drop of 0.24 V in each conductor) if the full-rated current of the wire is drawn.

This solution has ignored any change in resistance due to self-heating of the wire which would increase the voltage drop.

Cables

Cables are used mainly for signal and data transmission and for interconnecting sub-systems within an electronic system. Table 2.2 summarizes the main types of

Table 2.2 Common cable types

Type	Construction	Applications
Screened	1 or more wires with an overall metal braided or helical screen and insulation	Low-power signal transmission at up to audio frequencies
Twisted pair	2 wires insulated and twisted together covered with overall sheath, possibly screened	Signal transmission at up to 10 MHz, RS 422 data transmission
Coaxial	1 solid or stranded conductor surrounded by dielectric, metal braid and outer insulation	Signal transmission at up to 1 GHz
Twin feeder	2 conductors laid parallel about 10 mm apart, insulated and separated by a web of plastic	Radio receiver antenna downleads
Ribbon	10–50 stranded conductors laid parallel and coplanar covered and separated by insulation	Parallel logic interconnection in microprocessor and computer systems

cable used in electronic engineering. Multi-core cables can combine these types within one cable for special applications. Each type of cable has its own range of uses and its own characteristic parameters.

There are only a few applications in electronic engineering for cables of the type known as *flex*. These consist of several insulated wires with overall insulation, such as three-core mains flex used to connect mains-powered equipment to a mains outlet (this is one of the few applications). The reason for this is that cables are designed to carry electromagnetic signals and the cable must either exclude unwanted signals present in the surroundings (interference) or else prevent energy escaping from the cable (and causing interference elsewhere). A screened cable is intended to prevent pickup of unwanted signals. The wire or wires within the cable carry the signal (typically from a microphone or other transducer) and are surrounded by a metal screen wound helically or woven from bare wires in the form of a braid. The screening is effective only against electric fields and high-impedance electromagnetic fields (with a strong electric component). Magnetic fields and low-impedance electromagnetic fields (with a strong magnetic component) cannot easily be screened against. The effect of a magnetic field on a cable can be reduced, however, by twisting a pair of wires together. This means that currents induced in the wires by a changing magnetic field tend to cancel because each twist of the wires reverses the polarity of the wires relative to the field. A twisted-pair cable can be used to transmit frequencies of up to 10 MHz, but energy will be radiated at megahertz frequencies and a coaxial cable should properly be used for the higher frequencies. Coaxial cables will carry signals down to zero frequency, but their main use is for transmission of radio-frequency (r.f.) signals at up to 1 GHz.

Chapter 7 discusses electromagnetic effects in greater detail. Carter discusses electrostatic and magnetic screening.

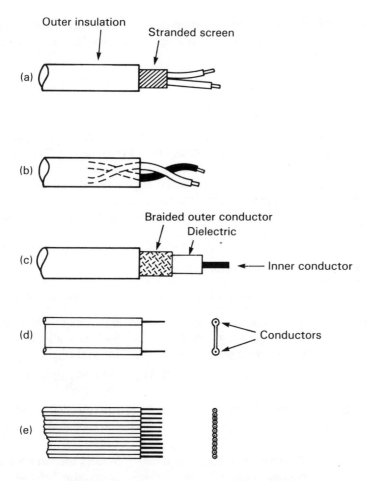

Fig. 2.5 Common cable types: (a) screened; (b) twisted pair; (c) coaxial; (d) twin feeder; (e) ribbon (insulation displacement).

Any cable which is longer than the wavelength of the signals being carried must be regarded as a *transmission line*. Cables designed to carry signals of frequency higher than audio frequencies therefore have characteristics which include transmission-line parameters. The two most important characteristics are the characteristic impedance, Z_0, which is typically 50–150 Ω and the attenuation, α, which is usually stated in dB per metre, 100 m or km (the frequency must also be given since α depends on frequency). Three other parameters commonly stated for a cable are the capacitance per metre (typically < 100 pF) the maximum working voltage, and the operating temperature range.

Ribbon cables are widely used in microprocessor and computer systems to carry parallel logic signals over distances of up to 5–10 m. Their main advantage over conventional multi-core cables is that they can be mass terminated: a connector can be fitted to the cable in one operation taking less than a minute. If a ribbon cable is carrying high-speed digital signals, each conductor should operate as a transmission line. For this reason it is common practice to earth alternate conductors. If an even number of signals, n, is to be carried, as is normally the case,

The theory of transmission lines is covered by Carter.

the cable should have $2n + 1$ conductors so that all signal conductors have an earth on both sides. This ensures that the characteristic impedances of all the signal conductors are equal.

Connectors

Connectors, or plugs and sockets, are used in electronic products to make electrical connections which can be easily disconnected and reconnected. They are used to connect external cables to equipment and are also fitted internally to allow sub-assemblies (such as PCBs) to be disconnected and removed easily for repair. One can therefore classify connector types into two main categories: cable connectors, suitable for making external connections to equipment; and wiring connectors, designed for use internally.

Table 2.3   Common types of cable connectors

Type	Construction	Features/applications
BS 4491 CEE 22	3-pole male body with female contacts, female sockets with male contacts	6 A rating mains connector, sockets available with integral filters
DIN[1] audio	Metal shell with 2–8 contacts	Domestic quality audio equipment
BNC 50/75 Ω	Coaxial bayonet	Instruments, general screened and coaxial connector
'D' type	9–50 way connectors with contacts in 2 or 3 rows available for soldering, wire-wrapping ribbon cabling and PCB mounting	Widely used multi-pole connectors, 25-way version used for 'RS-232' data transmission, 37-way for 'RS-449'

[1] Deutsches Institut fur Normung (the equivalent of the British Standards Institution in the Federal Republic of Germany)

Table 2.3 lists a few common types of cable connector and is limited to widely used, standardized designs. There are many types of proprietary connector, especially for use with multi-cored cable. Most cable connector types are keyed or polarized mechanically in some way so that they can be connected in one position only. Connectors designed to be attached to the end of a cable are called *free* connectors and incorporate some form of strain relief to grip the cable mechanically so that tension in the cable is not transmitted to the electrical joints inside the connector. *Fixed* connectors are designed for mounting on a panel or PCB.

The contacts within a connector are referred to as male or female. When two mated connectors are separated the live side of each circuit, if any, should be on the female contacts. The male contacts (which are accessible) should be electrically dead. In nearly all cases, this is a positive safety requirement and is one of the consequences of the safety rule that live parts shall not be accessible.

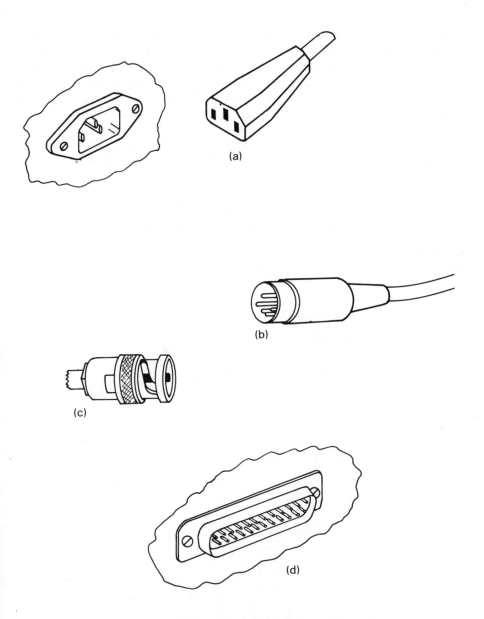

Fig. 2.6   Four common types of cable connector: (a) BS 4491/CEE 22 3-pole mains connector; (b) 5-pin DIN audio; (c) BNC 50 Ω coaxial; (d) 25-way male 'D'-type.

Some of the factors to be considered when choosing a connector are: the electrical ratings and characteristics such as maximum working voltages and currents, contact resistances, insulation resistance and transmission line parameters; the temperature ratings and intended operating environment; the reliability and life of the connector; cost; and tooling needs for making the electrical connections. The transmission line parameters are applicable only to radio-frequency or r.f. connectors and will include the characteristic impedance and the voltage standing wave ratio (VSWR).

When an electromagnetic wave travelling along a cable meets an electrical discontinuity such as a connector, some of the wave energy is reflected and sets up a standing wave. The VSWR is a measure of the amount of reflection from the discontinuity. Carter, Chapter 7, discusses VSWR and other transmission-line parameters.

The environmental conditions under which a connector will be working are most important. Connector types capable of use out of doors (splashproof or waterproof) are much more expensive than types intended for indoor use, because of the complexity of the waterproof seals and the need to protect the connector from atmospheric corrosion. The life of a connector is influenced by the number of mate/unmate operations: a heavy duty connector is designed to withstand the wear and tear of frequent use, while other types are designed for occasional mating and unmating only. Many cable and connector types require special tooling to prepare the end of a cable and to make the electrical connections to the connector. Additionally, some training and skill is needed if a good quality termination is to be made. A possible solution to this problem is to buy-in terminated cables from a specialist cabling contractor. Some common cable and connector configurations are available commercially as ready-made cable assemblies.

Wiring connectors are somewhat less standardized than cable connectors and different manufacturers' products are often not interchangeable. They are used for making connections to the edges of PCBs (edge connectors), for attaching flying

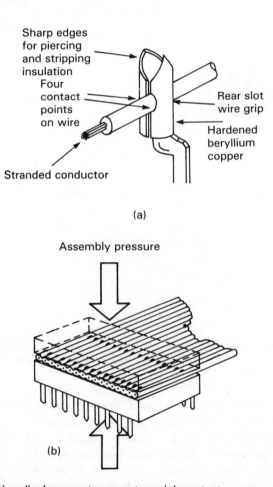

(a)

(b)

Fig. 2.7   Insulation displacement connectors: (a) contact arrangement; (b) DIL plug connector assembled to cable. (Courtesy Thomas & Betts Ltd. Design covered by US Patent 3.964.816).

leads to PCBs and for connecting ribbon cables. (Some ribbon cable connectors are suitable for making external connections, but in general their use is confined to interior interconnections.) Many wiring connectors are attached to wires by crimping rather than soldering, because crimping is a much quicker operation in production.

Ribbon cable connectors make electrical contact with the cable conductors by *insulation displacement*. Sharp forked prongs within the connector pierce the cable insulation and cut into the cable conductors as shown in Fig. 2.7. All the connections within the connector (up to 50) are made simultaneously in a single pressing operation, making this type of cable and connector an economical and virtually error-free method of interconnecting logic PCBs. Ribbon cable connectors are also available for connection to dual-in-line (DIL) integrated circuit sockets, as shown in Fig. 2.7(b).

Where individual wires have to be connected, push-on terminals (of the type commonly used in automobile wiring) or screw terminals are often used. Some screw terminals are specifically designed to accept bared wires (tinned with solder if multi-stranded), others are designed to take a spade or eyelet terminal crimped to the end of a wire.

Printed Circuits

Printed circuits are used in almost all application areas of electronic engineering. *Rigid* printed circuit boards account for the majority of applications, providing both mechanical mounting and electrical interconnection for components. *Flexible* printed circuits are becoming increasingly popular as a substitute for discrete wiring. A flexible circuit cannot provide adequate mechanical support for anything other than the smallest components. This has led to the development of the flexi-rigid board, fabricated in one piece with some sections flexible and some rigid.

Not all PCBs have electronic components mounted on them and a significant use for PCBs is to provide interconnections between other PCBs, particularly in *card cages* or *card racks* where a number of boards slide into a frame and connect with a *backplane* which provides electrical interconnection between the boards. Another possibility is to use a PCB as an electrical and mechanical base on which to mount sub-boards containing functional blocks of circuitry of standardized design. This is known as the *motherboard–daughterboard* technique.

An interesting development of this idea is described by John Ashman in 'Hierarchical Interconnection Technology', *Electronics and Power*, 671–673, **32**(9), September 1986.

Printed circuits are manufactured by chemical etching and electroplating processes. The patterns of conductors, or *tracks*, on a printed circuit are defined photographically from a master photographic film. The film itself is made either by direct photoplotting from a computer-aided design (CAD) system or by photographing an image of the conductor pattern or *artwork* prepared by a draughtsman.

Rigid Printed Circuit Boards

There are three types of rigid printed circuit board shown diagrammatically in Fig. 2.8. Single-sided boards with a conductor pattern on one side only are the cheapest type. They are mainly used for low-cost, low-component-density, consumer applications such as portable radios. Double-sided boards can carry a

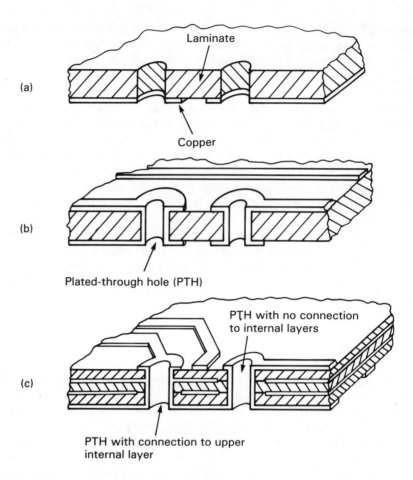

Fig. 2.8 PCB construction: (a) single-sided; (b) double-sided plated-through; (c) multi-layer.

greater density of circuit interconnections and are much more common than single-sided boards. Most double-sided boards manufactured have plated-through holes (PTHs) so that interconnections between one side of the board and the other are made during board manufacture. Through-hole plating is a process by which copper is deposited on the inside walls of holes drilled through the board. A thin layer of copper is first deposited by *electroless plating* onto the non-conducting board material. The copper thickness is then built up by electroplating. Double-sided boards without plated-through holes are cheaper to manufacture because fewer process steps are involved, but are more expensive to assemble in most cases because link wires or pins have to be hand-soldered through some of the holes to make connections between one side of the board and the other. Multi-layer boards have internal layers of conductors as well as the conductors on the outer faces of the board. Connection to the internal layers is by plated-through holes. The manufacturing process for multi-layer boards is more elaborate (and therefore more expensive) but makes possible higher-density boards than could be achieved otherwise. Multi-layer boards also offer better electrical performance, for reasons which are covered in Chapter 7.

Multi-layer boards can be fabricated with more than twenty conductor layers although less than ten is a more usual figure.

16

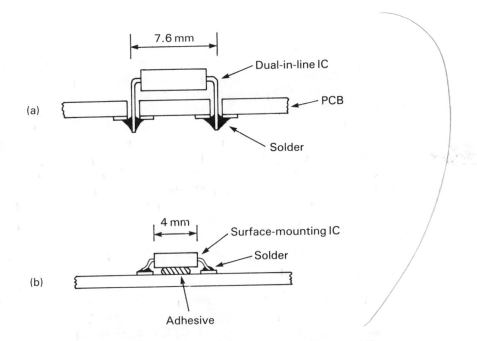

Fig. 2.9 Component mounting methods: (a) through-hole; (b) surface mount.

Component Mounting

There are two methods of mounting components on a PCB illustrated in Fig. 2.9. Through-hole mounting, in which component leads pass through holes in the board, is a technique which has been used from the advent of printed wiring. The alternative mounting method is to attach component leads or pads to the surface of a PCB without passing through the board. Surface mounting was tried during the 1960s, using welding to attach integrated circuit leads to copper, but has now become feasible using solder jointing. The main advantage of surface mounting is the smaller size of surface mount components, which allows greater component density on PCBs. Holes are still needed, of course, to make connections between one side of the board and the other, or to connect to internal layers, but these holes can be much smaller than those needed for component leads.

Board Materials

Most professional printed circuits are made from fibreglass-reinforced epoxy resin laminate. The fibreglass reinforcement is usually in the form of woven cloth. For low-cost applications (such as pocket transistor radios) synthetic-resin-bonded paper or s.r.b.p. may be used. Fibreglass boards are more heat resistant and more stable dimensionally than s.r.b.p. and also absorb less moisture from the air, resulting in better long-term insulation resistance. Flame-retardant grades of fibreglass board are available for applications where there is a risk of severe overheating. One advantage of s.r.b.p. is that holes may be punched rather than drilled, which reduces the manufacturing cost provided the expense of preparing the punch tool can be recovered over a large number of boards produced.

Manufacturing Processes

The starting point in the manufacturing process for single- and double-sided boards without plated-through holes is a photographic film, actual size, defining the pattern of conductors required on the board, and a piece of board material coated on one or both sides as appropriate with a layer of copper. The copper is then coated with *photo-resist* which is sensitive to ultraviolet light. The photographic film is then laid in contact with the resist and the whole assembly exposed to ultraviolet light. Exposed regions of the photo-resist undergo a photochemical change, leaving an imprint of the conductor pattern in the resist. The exposed board is then chemically *developed* in a tank of developer or in a conveyor machine in which developer is sprayed onto the board. This removes the resist, except in areas where copper will be left on the final board. These areas remain covered by resist – so called because its job is to protect these areas of copper from chemical etching. After development the pattern of conductors on the final board is visible as the pattern of resist.

The developed board is then etched, usually in a warm solution of ferric chloride, to dissolve the unwanted copper. After etching, the remaining resist is removed with a solvent and the board is ready for drilling.

Boards produced in quantity are drilled automatically on a numerically controlled (NC) drilling machine. Drilling coordinates for an NC drill can be generated automatically from a CAD system. Drilling may also be done by hand in a vertical drilling machine, by eye (sight drilling) for a prototype board, or using a jig for small batch production work.

The manufacturing process for double-sided PTH boards starts with drilling of the blank board. The drilled blank has copper all over both faces to conduct electroplating current to the holes in the board. The inside walls of the holes are therefore plated with copper before the track pattern is etched onto the board. During etching the etch resist protects the insides of the holes from the etchant. There are many variations on PCB manufacturing techniques which cannot be discussed here for lack of space.

Multi-layer board manufacture is a more elaborate process than double-sided PTH board manufacture but has some steps in common. The internal copper layers of a multi-layer board are etched individually as described above for conventional boards, except that each layer of board material is thinner than conventional board laminate. Intermediate layers without copper, known as *pre-preg*, are not fully cured: the epoxy resin has only been partially heat-treated and is still capable of plastic flow under heat and pressure. When all the internal copper layers have been etched, the sheets of laminate and pre-preg are assembled together and bonded under heat and pressure in a press. Normally, a stack of boards is pressed at the same time, separated from each other by sheets of steel and PTFE-based plastic film.

After bonding, the board is externally similar to a double-sided PTH board: the outer faces are still covered completely with copper. The board is drilled and the through holes are plated in the same way as a double-sided PTH board. With multi-layer boards, of course, some of the plated holes make electrical contact with internal copper layers so that the quality of the drilled holes has to be good. Finally, after through-hole plating the outer faces of the multi-layer board are etched and finished in the same way as double-sided PTH boards.

The thickness of the copper layer is stated in μm or as a weight per unit area (often ounces per square foot). A common thickness is 35 μm (1 ounce-foot^{-2}).

Scarlett describes many variants on the basic techniques discussed here.

18

Solder Resists and Legends

Most printed circuit boards are coated with an epoxy resin material called solder resist, usually dark green in colour. The purpose of this coating is to prevent solder sticking to unwanted areas of the board during mass-soldering, perhaps causing short circuits, or *solder bridges*, between adjacent tracks. It also serves two secondary purposes: less solder adheres to the board during soldering (solder is expensive) and the coating reduces moisture absorption during the board's life, thus improving reliability. The solder resist is applied as a liquid by *screen printing*, through a fine mesh screen.

The final manufacturing process, before assembly of components onto the board, is printing with a *legend*. This identifies the component positions and reference numbers on the board, the board type number and perhaps the date of manufacture or the version number. All of this information is useful if the board has to be repaired. The legend is screen printed onto the board on top of the solder resist with an epoxy-based ink.

Printed Circuit CAD

Most PCBs for digital and low-frequency analogue applications are now designed using CAD systems, and the films required for board manufacture are generated by photoplotting. Manual design and artwork preparation is still used for high-frequency and microwave boards.

The first decisions to be made in designing a PCB are the size and shape of the board, if not already determined by the application, and the method of construction to be used. Unless there are electrical reasons for choosing multi-layer construction, there is often a choice between multi-layer and double-sided plated-through boards. The time taken to design a board can run into weeks so that if a double-sided design has to be abandoned part-way through because of difficulty in accommodating all the connections a considerable amount of money and time will have been wasted. On the other hand, the choice of multi-layer construction adds additional material and manufacturing costs to every board produced.

When the board dimensions and outline have been decided, the second stage in the design process is to decide how and where to position components. The component placements will be influenced by the logical structure of the circuit, by electrical requirements such as power distribution and by the need to minimize interactions between different parts of the circuit.

Sensitive amplifier inputs, for example, would normally be positioned well away from power supply rails and output signals.

The final step in the design process is to find a path for every connection in the circuit. In practice, this stage of design interacts with component placement as some components may have to be rearranged. Finding a path, or route, for each connection is not easy as tracks cannot cross each other on the same side or layer. A common design for double-sided boards is to have all the tracks on one side of the board running horizontally and all those on the other side running vertically. Where a track changes from one side of the board to the other a *via hole* is used, not a component lead hole. The process of finding paths for the tracks is called *routing* and is often performed automatically on a CAD system. On a complex PCB, a CAD system may not be able to route all the tracks, leaving a few to be routed manually.

CAD systems incorporating software techniques from the field of artificial intelligence may be able to achieve full routing in all cases.

19

Preparation of Artwork

Unless the photographic films required for PCB manufacture are to be plotted directly from a CAD system on a photoplotter, photographic artwork will be needed. The artwork is, in effect, reproduced photographically on the final manufactured boards and must therefore be of high quality and accuracy. The traditional method of preparing artwork is manual taping. The artwork is prepared twice full size (2:1) or four times full size (4:1) by sticking self-adhesive black tape and sheet onto transparent drawing film. Common patterns of conductors such as transistor and integrated circuit pads are available commercially. For large areas, sheet material is used, cut with scissors to the required shape. To make the draughtsman's work easier the drawing sheet is laid onto a light box and illuminated from below. Artwork for solder resists or earth planes which covers more of the board than it leaves exposed is normally prepared in negative with black tape or pads defining the areas which will be left bare on the final board. Conversion to a positive is done during photographic reproduction.

Once the artwork is complete, it must be checked thoroughly against the circuit diagrams to ensure that all connections have been included and that all hole spacings are correct for the components to be used. If any errors slip through the check a fresh set of photographic films will have to be prepared after the artwork is corrected. The checked artwork is then photographically reduced (accurately) to final size and a set of working films prepared from the master films.

One problem with taped artwork is storage: the sheets must be carefully stored and handled. It is not uncommon for the tape to peel off after some time, so that if the artwork is modified and re-photographed it may also have to be completely rechecked. Photoplotted films or plotted artwork produced by a CAD system do not suffer from this problem.

Printed Circuit Assembly

Mass production of assembled and soldered PCBs is highly automated. Components are inserted into a board by high-speed automatic component insertion machines, or in the case of surface-mounted components by machines which fix the components to the board with a drop of adhesive. Some components may have to be inserted by hand, but this is becoming less common as more components are designed for automatic handling. Once all the components are in place, the whole board is mass-soldered, so that all soldered joints are made in one fast, cheap process. There are two main mass-soldering techniques in use: wave soldering and reflow soldering.

Wave-soldering

Mass-soldering of conventional through-hole printed circuit boards (as opposed to surface-mount boards) is normally done in a wave-soldering machine. Figure 2.10 illustrates the principles of the process. PCBs loaded with components pass along a conveyor over a wave of molten solder maintained by a pump from a solder bath. The underside of the board is preheated and fluxed before the board reaches the wave. As the board passes across the wave the underside of the board is washed with molten solder. If conditions are correctly adjusted, just sufficient solder stays

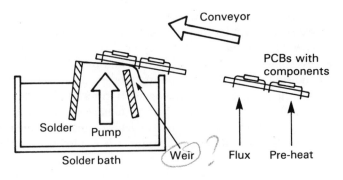

Fig. 2.10 Principle of wave-soldering.

on the board to make good joints, with no globules or icicles of solder on the component leads which are later cut off with rotating cutters.

Reflow Soldering

Surface mount PCBs are normally mass-soldered by a reflow process. Components are stuck to the board with adhesive and a solder–flux paste is applied to the joints. The board is then heated to melt or reflow the solder by one of two methods. In *vapour-phase reflow* the board is passed through a tank in which an inert fluorocarbon liquid is boiling, filling the tank with hot vapour. The vapour condenses on the board, giving up its latent heat of vaporization to the board and thus heating the solder joints. Precise temperature control is achieved because the board is heated to no more than the boiling point of the liquid. In *infra-red reflow* the solder paste is heated by infra-red radiation from electric elements.

Design Considerations

Manufacture of a board can be made easier and cheaper by good PCB design. If a board is to be wave soldered for example, the direction of flow of the solder should be checked before the board is designed and tracks on the solder side of the board laid out in the direction of flow. This reduces the chance of solder forming bridges between tracks. For the same reason, ICs should be laid out with their rows of pins across the direction of flow. If a board is to be assembled with automatic component insertion equipment, setting-up of the machine is easier if all axial components are laid out on a common pitch and all integrated circuits are oriented in the same direction.

Further recommendations on PCB design are given in BS6221: Part 3: 1984.

Rework and Repair

The ease with which wire-wrapped joints can be remade is a distinctive feature of wire-wrap technology. Soldered connections on PCBs and elsewhere may also need to be altered or remade for several reasons.

Minor faults can occur in PCB manufacturing and assembly which require some manual attention to the board. The process of correcting production faults is called *rework*. If a board needs attention to correct a fault which has occurred in service the work is called *repair*. Similar manual techniques are applicable in either case, although the types of fault encountered may be different. Rework may involve removal of excess solder from a board, removal and replacement of an incorrect component or addition of a component omitted during manufacture. Repair may involve reconnection of a broken wire or PCB track, replacement of failed components and restoration of a mechanically damaged or burnt area of PCB. Some types of repair and rework on PCBs are delicate jobs for skilled craftsmen, but some limited skill at component replacement is needed by most electronics engineers.

One of the most common rework and repair operations is removal of solder to release a component from a PCB. There are two main ways of doing this, one using capillary action and one using partial vacuum. Copper braid impregnated with flux can be used to draw the solder out of a joint or hole by applying a soldering iron to the braid while in contact with the solder to be removed. Once the solder is molten, capillary action draws it into the braid and out of the joint. The end of the braid is then cut off and discarded. An alternative, less messy, technique is to use a solder-sucker syringe or suction soldering iron. Here, the iron is used to melt the solder and a sharp suck from the spring-loaded syringe or suction iron removes the molten solder from the joint. Both techniques require skill if the joint is not to be over-heated, causing damage to nearby components or delamination of a PCB track. Component removal on plated-through boards can be especially difficult: when removing an IC it is better to cut all the IC leads to release the body and then unsolder the leads one by one. It is fairly easy to pull out the barrel of a plated-through hole while removing a component lead. On a double-sided board this may not be a problem as the new component can be soldered on both sides of the board, but if the hole connects to an internal layer on a multi-layer board further repair may be impossible.

Damaged tracks on a PCB can be repaired by soldering discrete wires to the board to bridge across a damaged area, or by using self-adhesive copper strip which can be stuck to the board and soldered to the undamaged parts of the track. If alterations are needed to a PCB, the same methods can be used: tracks can be disconnected by cutting across in two places with a sharp knife and then carefully lifting the piece in between with the end of the blade.

Damaged areas on a PCB can be restored using epoxy resin after removal of any loose fragments or burnt areas (caused for example, by a component being burnt out by a fault).

Further details of repair and rework techniques can be found in BS6221: Part 21: 1984.

Case Study: A Temperature Controller

There is no automatic method of assembling and soldering discrete wires. Wire-wrapped wiring can, however, be connected by machine in applications such as computer backplanes.

This chapter has introduced several interconnection technologies including discrete wiring, printed circuits and connectors. Modern electronic products are designed without discrete wiring as far as is possible because of the high cost of assembling and soldering wires. There are several different applications for discrete wiring in electronic products, summarized in Table 2.4 alongside alternative interconnection techniques which can be used for the same applications, but which avoid the need to solder wires by hand.

The product illustrated in Figs 2.11 and 2.12 is a temperature controller,

Table 2.4   Alternatives to discrete wiring

Application	Alternative techniques
Wiring to front panel controls, connectors and displays.	Direct PCB mounting switches, potentiometers and connectors. PCB fitted behind panel with ribbon cable interconnect to main PCB or right-angle mounting components fitted direct to main PCB.
Connection to transformers and power supplies.	PCB mounting components. Push-on crimped wiring connectors with PCB terminals soldered into board during mass soldering.
Inter-PCB wiring where a product consists of more than one PCB	PCB motherboard. Board-to-board connectors. Ribbon cables connecting to PCB sockets. Flexi-rigid boards made in one piece and folded to fit into the product.

manufactured in several versions with different types of input and output. Figure 2.11 shows an early version of the controller while Fig. 2.12 shows a later Mark 2 version. Even more recent designs use digital displays and push-button controls for setting the temperature and are also designed from the start to have no discrete

Fig. 2.11   Mark 1 temperature controller. (Product illustrated courtesy Eurotherm Ltd)

Fig. 2.12   Mark 2 temperature controller. (Product illustrated courtesy Eurotherm Ltd)

wiring and to be easily assembled, so they do not show the advantages of modern interconnection techniques as well as the early design illustrated.

The Mark 1 version of the controller, shown in Fig. 2.11, has two PCBs. The larger board includes the input and output terminals at the rear edge, a transformer and power supply, temperature control circuits and the main output circuit. The smaller board contains circuitry which varies from one version of the product to another to accommodate different types of optional second output circuit. The two PCBs are connected by a wiring loom of eight colour-coded wires, making sixteen joints to be soldered by hand. The wiring loom was assembled separately before soldering to the PCBs. A typical input to the controller is from a thermocouple sensing the temperature to be controlled, while the output may be a thyristor circuit for controlling the power delivered to an electric furnace element. The desired temperature is set on the large wheel, viewed through a window in the front panel. The controller is shown without its outer cover, and is typically mounted into an industrial control panel.

The Mark 2 version, shown in Fig. 2.12, has no discrete wiring. The smaller PCB is fabricated with edge contacts which locate onto square pins staked into the main PCB which are mass soldered with all other connections on the main board by wave soldering. Four connections at the rear of the smaller board are made with a wiring connector pushed onto pins in the main PCB. There is thus no hand soldering needed and the wiring loom has been eliminated. The connections formerly made by the wiring loom are now made at low cost by wave soldering with no possibility of error. This means that there is no need for the connections to be checked: a visual inspection of the solder joint quality is sufficient. The change in wiring

The construction method used is an example of the motherboard technique discussed earlier in this chapter.

design together with other design changes means that the Mark 2 version can be assembled by only five people compared to seven for the Mark 1. These five people assemble 800 units per week. More recent designs are assembled *and tested* by only three people at the rate of over 1000 per week.

The objective of modern interconnection techniques in electronic product design is to minimize *total* manufacturing costs and this involves a balance between material and parts costs, assembly costs and capital costs for assembly equipment. Other factors such as parts inventories and the amount of work in progress on a production line are also important. Many of these aspects of electronic production depend on product design and *design for production* is an important objective in modern electronic engineering design.

Summary

Interconnection technology is of fundamental importance in the design and manu-facture of electronic products. Electrical joints in electronic systems are most often made by soldering although for some purposes, wire-wrapping or crimping is used. Soldered joints are made with a low-melting-point tin–lead alloy which dissolves into the surfaces of the metals being joined. Cleanliness, flux and sufficient heat are essential requirements for a good soldered joint. Different solders are available for different applications: the type of solder to be used should be selected with care. Wire-wrap jointing is an alternative to soldering for some applications and has advantages in ease of alteration for prototype work. The integrity of a wire-wrap joint depends on good metal-to-metal contact brought about by the pressure of the wire on the sharp corners of the terminal pin.

Discrete wiring and cabling is best avoided where possible, but is still essential in many applications. Wiring is used mainly for making internal connections and cabling for external connections. The selection of wires and cables for a particular application requires care and attention to the characteristics of the wire or cable. Connectors are used where a connection must be easily disconnected and reconnected, and for connecting cables to equipment. The selection of a connector requires the same care as the selection of any other electronic component.

The principal interconnection technology described in this chapter has been the printed circuit, manufactured by chemical etching using photographic techniques to transfer a conductor pattern from a film to a board. Single-sided PCBs are the simplest and cheapest type. Double-sided and multi-layer boards with connections between sides and to internal layers by plated-through holes are manufactured by more elaborate processes and are therefore more expensive. Printed circuit designs can be prepared on a CAD system or manually using taped artwork. PCBs can be mass-soldered using a wave-soldering machine or by vapour-phase reflow.

Alterations and repairs to PCBs require skilled techniques and some special tools for removal of solder from joints.

Problems

2.1 Calculate the minimum cross-sectional area of copper wire required to carry a current of 10 A with a voltage drop of less than 50 mV per metre of

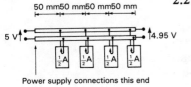

50 mm 50 mm 50 mm 50 mm

5 V

4.95 V

$\frac{1}{2}$A $\frac{1}{2}$A $\frac{1}{2}$A $\frac{1}{2}$A

Power supply connections this end

conductor. The resistivity of copper is 1.7×10^{-8} Ω m. Ignore any heating effects.

2.2 A printed-circuit backplane is to be designed for a microcomputer. Four logic boards are to be plugged into the backplane, each drawing $\frac{1}{2}$ A at 5 V and spaced 50 mm apart. The 5 V and 0 V power-supply connections are at one end of the backplane 50 mm from the first board. What width of track is needed in 35 μm copper if the voltage at the far end of the backplane is to be no less than 4.95 V? The resistivity of copper is 1.7×10^{-8} Ω m. Ignore any heating effects.

Integrated Circuits

3

Objectives

☐ To emphasize the importance of integrated-circuit technology in electronic engineering.

☐ To describe the manufacturing processes used in integrated-circuit production from preparation of raw material to packaging and testing.

☐ To introduce handling precautions for semiconductor devices.

☐ To outline the range of custom integrated-circuit design methods from full-custom through gate arrays to programmable logic.

☐ To discuss briefly the importance and pitfalls of second sourcing.

The widespread application of modern electronic products is made possible, above all else, by the technology of the monolithic integrated circuit (IC), first developed during the 1960s. Integrated circuits have had a profound effect on electronic product design and utility in at least four ways. Firstly, IC technology is inherently a mass-production technology, producing low-cost devices. Secondly, ICs are miniaturized circuits typically less than about 8 mm across. This makes it possible to manufacture small electronic products such as wrist-watches and pocket calculators which combine a high level of functionality with small size. Thirdly, ICs are reliable: good quality pocket calculators, for example, are more likely to fail through wear and tear on their keys than through the failure of their ICs. Compare this with the earliest computers built from thermionic valves which required several failed valves to be replaced *per day*. Finally, the advantages just described created pressure on IC designers to reduce the power consumption of their circuits, resulting in electronic products that can be powered by small primary batteries.

Monolithic means fabricated from one piece of (literally) stone. The development of monolithic IC technology was driven by the demands of the USA's Apollo space programme to land men on the Moon.

The same technology that produces the IC is also used to manufacture modern *discrete* semiconductor devices including transistors and thyristors. Externally, these components are simple, with three or four terminals and comparatively straightforward function compared to an IC. Internally, their detailed structure can be quite elaborate.

From the late 1960s onwards, the complexity of ICs as measured, for example, by the number of transistors on one chip, grew exponentially. Figure 3.1 illustrates this by plotting the complexity of one manufacturer's microprocessor chips on a *logarithmic scale* against year of introduction. As can be seen, the number of transistors per chip doubled roughly every eighteen months. Towards the mid-1980s, however, this trend began to slow down as IC manufacturers encountered the practical limits of then current IC fabrication technology. There were two contributions to the rapid doubling of IC complexity. One was a progressive reduction in the size of individual elements on the chip, the other was a gradual increase in the area of chips as improvements in process quality reduced the probability of defects on a chip. Innovation in circuit design also had some influence on chip complexity during the earlier years. The reduction in size of IC elements, such as transistors, caused several problems which have tended to slow

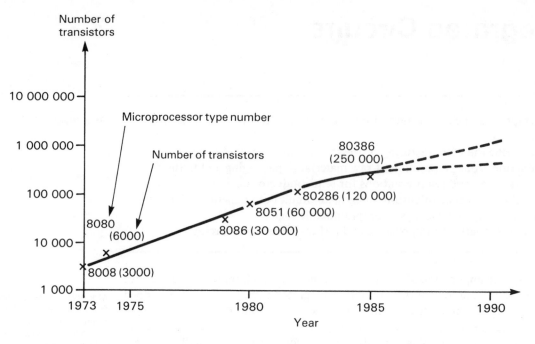

Fig. 3.1 Growth of IC complexity: number of transistors on microprocessor chips plotted on a *logarithmic* scale against year of introduction. (Source: Intel Corporation)

Devices may be scaled down to a few μm using simple calculations, but below this size other effects become significant and new circuit techniques are needed.

down the doubling of IC complexity. One was that small circuit elements required narrower lines to be defined on the chip during fabrication and the optical photo-lithographic methods and materials used are limited to linewidths of no less than about 1 μm. The electronic behaviour of circuit elements or *devices* is also affected by size reduction or *scaling*, so that changes in circuit design are needed as devices become smaller. The power density within a chip also increases as more and more devices are packed onto a chip and removal of heat becomes a problem.

LSI and VLSI ICs tend to be digital or logic circuits rather than analogue circuits because, with only a few exceptions, analogue circuits of high complexity are not needed. Because of the complexity of LSI and VLSI ICs they can only be designed with the aid of computers, or CAD.

The earliest ICs contained only a few tens of transistors or less than twelve logic gates and are now known as small-scale integration or SSI circuits. Later circuits, before the development of microprocessors, are known as medium-scale or MSI circuits and contain up to one hundred logic gates or several hundred transistors. Larger chips such as 8-bit microprocessors are known as large-scale or LSI circuits. Chips with more than 10 000 gates, such as 16-bit and 32-bit microprocessors are referred to as VLSI circuits, for very large-scale integration. Nothing larger is yet in commercial production, but the possibility of fabricating a monolithic circuit covering a whole wafer, from which chips are normally separated, has been suggested and given the name WSI, for wafer-scale integration.

The discussion in this chapter is mainly in terms of silicon because this element accounts for the majority of ICs produced today. A detailed discussion of semiconductor physics and the theory of semiconductor devices is outside the scope of this book. Till and Luxon give a more detailed treatment of most topics in this chapter.

Review of Semiconductor Theory

Pure (intrinsic) silicon has an electrical resistivity of 2.3×10^{-3} Ω m at room temperature, which is about eleven orders of magnitude greater than that of a good conductor such as copper at 1.7×10^{-8} Ω m and at least twelve orders of magnitude less than that of a good insulator such as polystyrene at about

10^{15}–10^{19} Ω m. Silicon is a tetravalent element in Group IV of the Periodic Table and has four electrons available for chemical bonding. Crystalline silicon is covalently bonded and has a tetrahedral lattice structure like that of diamond. The properties of silicon devices and integrated circuits depend on the ability of a silicon lattice to incorporate trivalent and pentavalent *dopant* atoms from Groups III and V of the Periodic Table. Group III dopants, such as boron, can contribute only three bonding electrons to the silicon lattice, leaving a vacant covalent bond or *hole*, which behaves as if it were a positive mobile charge carrier. Group III dopants are known as *acceptors* because the vacant bond accepts an electron from a neighbouring atom in the lattice. Silicon with an excess of acceptors is known as *p-type* silicon. Group V dopants, such as phosphorus and arsenic, have five bonding electrons, of which only four can contribute to bonding with the surrounding silicon atoms. The fifth electron is thus relatively free to move around the lattice and contribute to conduction. Group V dopants are known as *donors* because of this addition of a mobile electron to the lattice. Silicon with an excess of donors is known as *n-type* silicon. Doped silicon is known as an *extrinsic* semiconductor because its electrical properties are dominated by the effect of the dopant atoms.

Two atoms bonded covalently each contribute one electron to the bond. Bonding is discussed by Anderson *et al.*

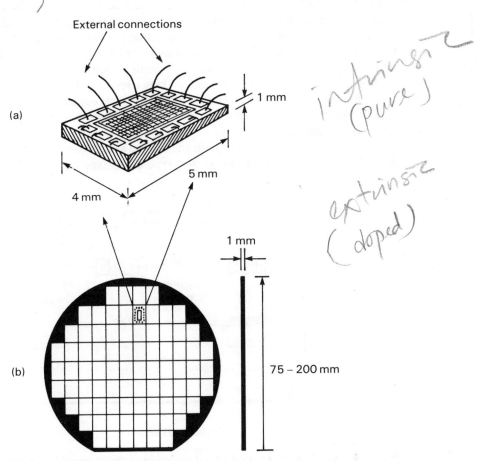

Fig. 3.2 Monolithic integrated circuits: (a) chip; (b) wafer from which chips are separated.

Structure of a Monolithic Integrated Circuit

Resistors and capacitors occupy more chip area than transistors. IC designers, therefore, tend to use circuit techniques which avoid the need for passive components. Ritchie gives examples of these techniques.

In practice diodes are often formed from a transistor with the base and collector connected together to form a *diode-connected transistor*, which requires less chip area by a factor of β than a diode of equivalent current rating. Ritchie gives examples of integrated-circuit designs using this technique.

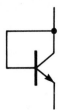

A typical IC chip is about 5 mm across by 1 mm thick, and has been cut from a *wafer* of semiconductor material containing hundreds of chips. Components within the integrated circuit such as transistors, diodes, and to a limited extent, resistors and capacitors, are built from regions of extrinsic semiconductor formed in the top 10–20 μm of the chip by incorporation of dopant atoms into the silicon crystal. Figure 3.3 shows how a p–n junction diode and an n–p–n junction transistor can be formed. The bulk of the chip shown has been doped to form a p-type *substrate.* An electrical connection to the substrate is made at the base of the chip for reasons which will become clear in a moment. Regions of n-type semiconductor have been formed by incorporating sufficient dopant atoms to create an *excess* of mobile electrons over the holes created by the p-type dopants. Within the n-type regions, further p-type and n-type regions have been formed by adding further dopants. The diode and transistor structures can be seen from the diagram to correspond with the conceptual structures shown. There are also some other p–n junctions present in the integrated circuit, formed by the n-type collector region of the transistor and the p-type substrate, and by the n-type region around the diode and the substrate. To maintain electrical isolation between the diode and the transistor, the substrate must be tied to the most negative potential in the external circuit so that the collector-to-substrate junction is reverse-biased. The surface of the chip is covered with silicon dioxide, which is an electrical insulator, and connections to the terminals of the diode and transistor are formed from aluminium deposited over the silicon dioxide, and making contact with the underlying devices through holes, or *windows,* in the insulator. External connections to the chip are made by fine gold wires connected to bonding pads around the edge of the chip using techniques described later.

There are other methods of isolating devices on a chip from each other and many possible geometrical arrangements of n-type and p-type regions, silicon dioxide

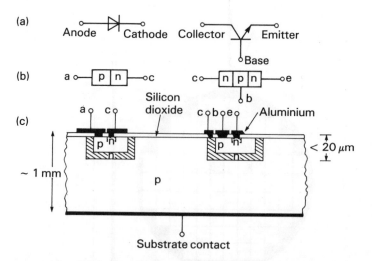

Fig. 3.3 Monolithic integrated circuit structure showing possible structures for p–n junction diode and n–p–n junction transistor: (a) circuit symbols; (b) conceptual structures; (c) integrated-circuit realization.

layers and metal layers. The geometry of IC devices and the sequence of fabrication steps required to create them are known collectively as a *process*. Development of a process can be a lengthy and expensive undertaking and precise details of proprietary processes are not released by manufacturers. There are, however, only a few major steps in a fabrication process, which are combined and repeated in various sequences to build up the desired IC structure. These process steps consist of *doping processes* to incorporate dopant atoms into the silicon lattice to form regions of n-type and p-type semiconductors of defined depth, lateral geometry and dopant concentration; *crystal growth* to build up new layers of silicon; *oxide growth* to form silicon dioxide layers; *lithography* to transfer images to the silicon; and *etching* to remove regions of silicon or silicon dioxide. The entire volume of an IC chip must be a single crystal of silicon, so the dopant atoms must either be incorporated into an existing crystal lattice or be included as the lattice grows.

Integrated-circuit Fabrication

IC fabrication is a mass-production activity. Dozens of wafers are processed simultaneously through many of the fabrication processes and each wafer contains hundreds of chips. Much of the processing is carried out at high temperatures of up to 1200°C, although there is a trend towards lower-temperature processing for the fabrication of ICs with small-geometry devices because of undesired dopant diffusion. The dimensions of individual device features on an IC are about 2–3 μm in current production ICs and down to around 1 μm on research designs. Devices are about 15–20 μm square in production chips.

Estimate the size of a storage cell on a type 27128 128 k bit EPROM chip which measures 4 mm by 5 mm. (1 k = 2^{10} = 1024).

Worked Example 3.1

Solution The chip area is 20 mm². Dividing this by 128 × 1024 the area of a storage cell is 153 μm². Assuming the cells to be square, they are about 12 μm across.

Very high standards of cleanliness are required in an IC fabrication facility, because dust particles are much larger than the dimensions of device features and the image of a dust particle reproduced photolithographically on an IC may obliterate several devices, rendering the chip useless. A very high standard of air filtration and operator cleanliness is therefore essential. Operators must wear special hooded overalls to reduce contamination with skin particles. Smoking is not permitted, even during breaks outside the clean room, because of the quantity of particles exhaled for some time after smoking. Cosmetics are also banned because of the danger of particulate contamination. There is a trend towards automation in modern IC fabrication facilities to reduce the number of operators required and therefore improve cleanliness.

   IC fabrication facilities are very expensive to set up and expensive to operate so they must produce large quantities of ICs, or ICs of high value, to be economically worthwhile.

Clean room air quality is defined by BS 5295 in terms of the number of particles of diameter greater than 0.5 μm per cubic metre of air. The figures given in BS 5295 are derived from an American standard based on particles per cubic foot, from which the designations Class 100, Class 1000 and so on are derived. (Class 100 = 100 particles foot^{-3}.) Humans typically shed 250 000 0.5 μm particles min^{-1} even at rest (over the whole body) and over 1 000 000 when moving about.

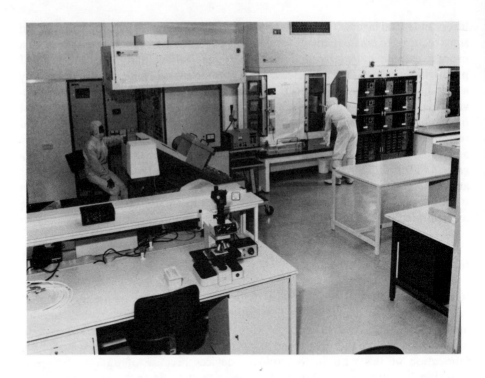

Fig. 3.4   An integrated circuit fabrication facility. (Courtesy British Aerospace)

Preparation of Silicon Wafers

Silicon occurs naturally as the second most abundant element after oxygen in the Earth's crust, making up about 25% by weight of the crust in the form of silicates and silica (SiO_2). Silica is found in a variety of forms including flint and quartz. Naturally occurring high-purity silica sand is the starting material for the preparation of device-grade silicon for IC manufacture. Silica is reduced to silicon by reaction with carbon in an electric furnace giving silicon and carbon monoxide:

$$SiO_2 + 2C \rightarrow Si + 2CO$$

The impure silicon is then converted to trichlorosilane ($SiHCl_3$) by reaction with hydrochloric acid

$$Si + 3HCl \rightarrow SiHCl_3 + H_2$$

and the trichlorosilane is then purified by distillation and converted back to poly-crystalline silicon by reaction with hydrogen. For device-grade silicon, typical residual impurity concentrations for Group III and V elements are less than 1 part in 10^9. Single-crystal silicon is required for IC fabrication. There are two methods of fabricating single-crystal ingots of silicon: the float-zone process and the Czochralski process. In the float-zone process a polycrystalline ingot is converted to single-crystal form by heating a small zone using radio-frequency heating. The molten zone is passed up the ingot and single-crystal silicon forms behind the molten zone. The float-zone process can also be used as a *refinement* or

Trichlorosilane is a colourless liquid with a boiling point of 33 °C at a pressure of 758 mm of mercury. The distillation is carried out at reduced pressure because trichlorosilane is unstable and cannot be distilled at atmospheric pressure.

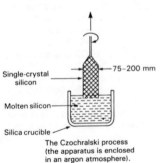

Single-crystal silicon
75–200 mm
Molten silicon
Silica crucible

The Czochralski process (the apparatus is enclosed in an argon atmosphere).

purification technique because impurities tend to remain in the molten zone. Most silicon for IC manufacture is made by the Czochralski process in which a single crystal ingot is formed by slowly withdrawing a rotating seed crystal from a crucible of molten purified silicon. The crystal orientation of the silicon is determined by the orientation of the seed crystal and is carefully chosen and indicated by grinding a flat along one side of the ingot. The crystal orientation of the wafers cut from the ingot is important when the wafers are scribed and broken into chips as cleavage occurs more easily along some crystal planes.

Ingots of up to 200 mm diameter can be grown. They are sawn into thin discs or wafers with a diamond saw. Small wafers of 75–100 mm diameter are about 0.5–1 mm thick. Larger sizes have to be thicker to prevent warping during processing. The surface of a sawn wafer is rough and damaged by the sawing process. The damaged layer is removed by lapping and the wafer is then chemically etched to leave an optically smooth mirror finish. The processed wafers are inspected for flatness because later processing depends on the projection of images onto the wafer surface, and any significant deviation from flatness will cause loss of definition in the image transferred to the wafer.

Till and Luxon discuss the importance of crystal orientation. For a general introduction to crystalline materials and crystal structure see Anderson *et al.*

Lapping is a surface finishing process using a fine abrasive paste.

Epitaxial Growth

Some semiconductor fabrication processes require the addition of extra layers of silicon on the surface of the wafer. Added silicon is known as an *epitaxial* layer and it must have the same crystal structure and orientation as the wafer itself and be grown as an extension of the wafer so that no crystal boundary exists between the original surface and the new layer. An epitaxial layer might be grown after creation of a doped region in the original substrate to form a *buried layer*, or to form regions of different doping type or concentration to the substrate. Epitaxial layers are typically less than 20 μm thick.

Epitaxial silicon can be grown on an *insulating* substrate with a suitable crystal structure. An important example is sapphire (Al_2O_3) used in the silicon-on-sapphire (SOS) process for CMOS circuits.

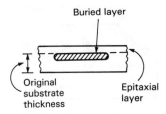

CMOS ?

How many silicon atoms make up a 20 μm layer?

Worked Example 3.2

Solution A crude approximation to within a factor of two or three can be calculated from the density and atomic weight of silicon, and the proton–neutron mass. (The mass of electrons in an atom is negligible.) The density of silicon is 2300 kg m^{-3}, the atomic weight is close to 28, and the masses of the proton and neutron are about 1.7×10^{-27} kg. A silicon atom has a mass, therefore, of $(28)1.7 \times 10^{-27}$ or 4.8×10^{-26} kg. The number of atoms in a cubic metre of silicon is $2300/4.8 \times 10^{-26}$ or 4.8×10^{28}. Assuming the atoms to be packed cubically (which they are not), there would be $\sqrt[3]{(4.8 \times 10^{28})}$ atoms along each edge of the cube. The number of atoms across the thickness of a 20 μm epitaxial layer is thus $\sqrt[3]{(4.8 \times 10^{28})} \times 20$ μm or about *70 000*, a surprisingly small number compared to the enormous numbers of atoms in bulk material.

Silicon atoms are added to the substrate surface in a reactor at a temperature of 800–1100 °C. The required silicon atoms can be produced by the pyrolytic decomposition of silane (SiH_4) at around 1000°C.

$$SiH_4 \rightarrow Si + 2H_2$$

or by reduction of silicon tetrachloride ($SiCl_4$) by hydrogen

$$2SiCl_4 + 2H_2 \rightarrow Si + SiCl_4 + 4HCl.$$

Dopants can be included as an integral part of an epitaxial layer by adding traces of gases such as diborane (B_2H_6), phosphine (PH_3) and arsine (AsH_3) to the gas flow through the reactor. Careful control of the gas concentrations is essential if the number of crystal defects in the epitaxial layer is to be minimized.

This is the overall reaction. The reaction

$$2SiCl_2 \rightarrow Si + SiCl_4$$

takes place on the silicon substrate after production of $SiCl_2$ in the gas stream by the reaction

$$SiCl_4 + H_2 \rightleftharpoons SiCl_2 + 2HCl$$

Oxide Growth

One of the most frequent steps in any IC fabrication process is the formation of a layer of silicon dioxide (SiO_2) on the wafer surface. Silicon dioxide is an excellent dielectric, and can be used as an insulating layer within a device such as an insulated-gate field effect transistor (IGFET) or to isolate an epitaxially grown region of semiconductor from the substrate. A final layer of silicon dioxide (apart from the metallization layer described later) can *passivate* the surface of an IC to protect it from atmospheric contaminants. The most significant application of silicon dioxide in IC fabrication is as a masking layer which is etched to define regions to be doped.

The thicknesses required for these two purposes are very different: typically 20 nm (0.02 μm) for an IGFET gate and up to 1 μm for an isolation layer.

A silicon dioxide layer can be produced by a chemical reaction between the wafer surface and either oxygen or steam or by a deposition process similar to epitaxial growth. Thermal oxidation requires a temperature of 800–1250°C controlled to within ± 0.5°C or better together with careful control of the oxygen concentration (by using, for example, oxygen–nitrogen mixtures) and the time of processing to within 5–20 seconds. Because oxidation is a reaction between oxygen and the silicon surface, the reaction rate reduces as the thickness of oxide increases. Growth of a 1 μm isolation layer can take over an hour, whereas the formation of an IGFET gate layer may take only a few minutes. The oxide layer forms at the expense of the underlying silicon because silicon atoms from the wafer react to form the oxide. The thickness of silicon converted to oxide is about 30% of the final thickness of the oxide layer.

Silicon nitride (SiN_3) is another possible dielectric used in IC fabrication.

Thermal oxidation is carried out in quartz furnace tubes slightly larger than the wafer diameter and up to 2 m long. The wafers are held vertically in quartz boats or carriers and are processed in large batches of perhaps a hundred wafers at a time.

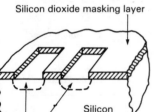

Silicon dioxide masking layer

Silicon

Regions to be doped

Lithography

Lithography means, literally, 'writing on stone'. The techniques discussed here are in principle, the same as those used for PCB manufacture and described in Chapter 2. The scale of lithographic images in IC fabrication is however about 500 times less than those used in PCB manufacture. (Typical *linewidths* are 2 μm and 1 mm respectively).

The various regions of extrinsic semiconductor, oxide and metallization making up the devices and interconnections on a chip have to be defined in the form of an image on the wafer surface. Techniques to do this are known as *lithography*.

The earliest technique used in IC fabrication and still very important today, is *photolithography*. Figure 3.5 illustrates the principle. The wafer is coated with a layer of *photo-resist* a few micrometres thick. The resist is applied to the wafer as a drop of liquid while the wafer is spinning at high speed, ensuring that the resist is

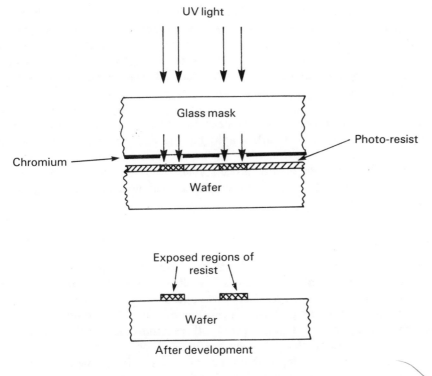

UV light

Glass mask

Chromium

Photo-resist

Wafer

Exposed regions of resist

Wafer

After development

Fig. 3.5   Principle of photolithography.

evenly distributed. The wafer is then gently baked to drive off the resist solvent. Resists are polymeric materials sensitive to ultraviolet light. Exposure to UV radiation can either cause a polymerization reaction or depolymerize the resist depending on whether a negative or positive image is required. The earliest resists for IC fabrication were of the negative image type, but for modern LSI and VLSI fabrication, positive resists are used because of their superior definition. After exposure the resist is developed in a chemical solution which dissolves the unpolymerized regions of resist, leaving selected regions of the wafer coated with a tough polymer to resist chemical etchants or ion beams.

The image pattern to be transferred to the resist is defined by a *photomask* or just *mask*. These are thin glass plates from 1.5 to 3 mm thick and as large as the wafer. The image is defined on the mask by a chromium layer which is generated litho-graphically from the IC design, either by photographic reduction from artwork or by the electron-beam lithography technique described below. The earliest ICs were fabricated by *contact printing*, in which the chromium side of the mask touched the resist-coated wafer. The obvious problem with contact printing was damage to the mask, and to overcome this, *projection printing* systems were developed so that the mask could be kept away from the wafer surface. Accurate alignment or *registration* is essential and both mask and wafer must have alignment marks to facilitate this. All the masks in a *set* for fabricating a particular IC design must of course be accurately made so that registration of the image defined by one mask with all others in the set is achieved.

Photolithography using ultraviolet light is limited to linewidths of no less than about 1 μm because of diffraction effects at line edges. Shorter-wavelength, X-ray, lithography can be used to overcome this problem, but at present is not used significantly in production. An important technique which also overcomes the resolution problem is *electron-beam lithography.* An electron beam of around 0.2 μm diameter is directed onto a resist-coated surface (either a mask or a wafer) on a high-precision *x–y* coordinate table. The resist must be an electron-beam resist, not an optical resist and the *x–y* table must be accurate to within fractions of a micrometre. The process is slow: full exposure of a 75 mm wafer can take more than an hour. Electron-beam lithography has the very significant advantage for mask fabrication of direct transfer of design information from a CAD system to the mask with no optical reduction process. It also finds application for fabrication of prototype chip designs without the expense of making a mask set, using *direct-write-on-wafer* imaging.

Etching

Etching is the removal of unwanted regions of material from a wafer. A typical example is cutting of holes or windows in a silicon dioxide layer prior to dopant diffusion or implantation into the regions under the windows. The regions to be etched are defined by the pattern in a resist created by a lithographic process as described in the previous section. There are three important etching methods used in IC fabrication: *wet etching; plasma etching; and ion milling.*

Wet etching was the earliest technique and is still used in production. Wafers to be etched are immersed in an acid bath and agitated to ensure even etching. Silicon is etched with a nitric/hydrofluoric acid mixture and silicon dioxide with a hydrofluoric acid/ammonium fluoride solution. The amount of material removed is dependent on temperature and immersion time.

Plasma etching is a dry technique in which reactive gaseous atoms react with the exposed regions of the wafer to form gaseous reaction products which are removed by a vacuum pump. The reactive atoms are generated by breakdown of molecules in a gas heated by radio-frequency electromagnetic energy.

Both wet and dry etching have the disadvantage that the etchant *undercuts* the resist as sketched in the margin. This effect has to be allowed for in the design of an IC layout and limits the packing density that can be achieved.

Ion milling is a physical rather than chemical etching process. The principle is illustrated in Fig. 3.6. High-energy ions are used to dislodge atoms from the region to be etched. The ions are accelerated electrostatically and formed into a collimated beam which is neutralized by the addition of electrons before reaching the target wafers. Because of the physical nature of this process, the resist is removed almost as fast, or faster than, the silicon or silicon dioxide being etched. The thickness of resist, therefore, needs to be greater than the depth of material to be removed.

Diffusion and Ion Implantation

In order to produce regions of n-type and p-type semiconductor within a wafer, dopant atoms must be introduced into the crystal structure. Dopants can be incorporated into epitaxial layers during deposition as described earlier. If dopants are to be incorporated into an existing crystal, they can be introduced by solid

A *plasma* is a fully ionized gas consisting of electrons and atomic nuclei. The term is used here to refer to a partly ionized gas consisting of electrons, ions and neutral atoms.

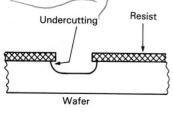

Undercutting   Resist

Wafer

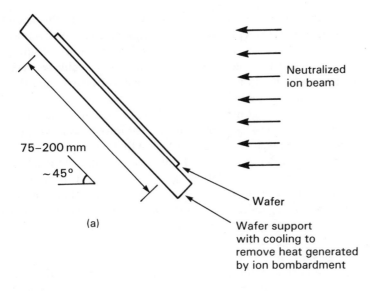

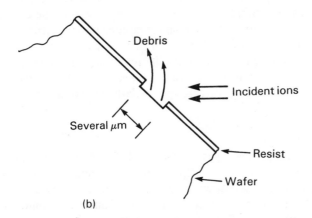

Fig. 3.6 Principle of ion milling: (a) large scale; (b) detail.

diffusion or ion implantation. Both techniques require an oxide masking layer to define the regions to be doped.

Diffusion was the earliest process used for doping a wafer and takes place in two stages: *predeposition* and the diffusion process itself. The dopant material can be deposited on the wafer by spin coating with a liquid or by deposition from a gas in a diffusion furnace. After deposition the dopant atoms are still concentrated near the wafer surface. The second stage of the diffusion process distributes the dopant atoms to the required depth in the wafer by heating the wafer to around 1000°C in a diffusion furnace, of identical construction to the furnace described earlier for oxide growth. Separate furnaces are essential for diffusion and oxide growth, otherwise the oxide furnace will become contaminated with dopants. During diffusion, the windows in the diffusion-masking oxide layer must be sealed to prevent the dopants from diffusing out of the wafer. This can be done by regrowing

Design rules specify the limitations of a fabrication process, including device sizes and minimum separations between doped regions. Modern IC CAD systems include programs called *design-rule checkers* which verify that all aspects of a design comply with the process design rules.

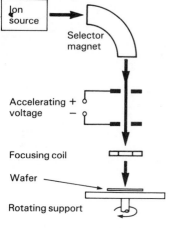

Ion source

Selector magnet

Accelerating + voltage −

Focusing coil

Wafer

Rotating support

oxide over the windows by using an oxidizing atmosphere in the diffusion furnace. Some predeposition processes produce a surface glassy layer which serves the same purpose. Careful control of diffusion time and temperature ensures that the dopant atoms diffuse to the desired depth, which is typically between 0.3 and several micrometres. Diffusion also occurs laterally, of course, and this must be allowed for in the process design rules.

Ion implantation is a newer method of doping in which the dopant is fired at the wafer as an ion beam inside a vacuum chamber. The ion energy is typically between 25 and 200 keV and can be precisely controlled so that the ions penetrate the wafer surface to a controlled depth. The total quantity of dopant introduced into the wafer can also be controlled to within $\pm 10\%$ or better. Typical ion doses are between 10^{15} and 10^{20} ions m^{-2}. The ion beam is broad compared to the line dimensions on the wafer, but is narrow compared to the wafer dimensions, so that the beam has to be scanned and the wafer rotated to ensure even exposure to the beam.

Implanted ions are not bonded into the silicon lattice: they occupy *interstitial* sites between the lattice atoms. The silicon lattice itself is also disrupted by the implanted ions as they lose energy by collision with the lattice atoms. After ion implantation, therefore, wafers must be heat-treated to repair the crystalline structure and activate the implanted dopant by incorporating the dopant atoms into the silicon lattice. Dopant atoms which are bonded into the lattice in place of a silicon atom are said to occupy *substitutional* sites. The heat-treatment process is known as *annealing* and is carried out at similar temperatures and for similar times as thermal oxidation and diffusion. During annealing, of course, diffusion occurs and the implanted dopants migrate through the lattice to some extent.

Metallization

Some interconnections may be made by diffused regions or by deposited polycrystalline silicon layers.

The final stage of wafer processing is deposition of a metal interconnect to connect together the individual devices on the chips and to form bonding pads for connection to the external circuit. Windows are etched in a final silicon dioxide layer to make contact with the separate devices. A layer of aluminium, sometimes with a small amount of silicon, is then deposited on the wafer, usually by vacuum deposition from aluminium vapour in a vacuum chamber. After deposition, the wafer is *sintered* at around 400°C. This improves the electrical connection between the deposited aluminium and the silicon.

A metal–semiconductor junction is known as a Schottky junction. Schottky transistors are used in Schottky and low-power Schottky TTL logic circuits.

In some devices, such as Schottky diodes and transistors, the metallization forms part of the device as well as being an interconnect. This is also true in metal-gate MOS devices, where the gate electrode is fabricated as part of the metallization layer.

Testing, Dicing and Bonding

Testing LSI and VLSI circuits is a non-trivial operation. Wilkins gives an introduction to the problem and discusses some current approaches to *testable* design.

After metallization all the individual ICs on the processed wafer have to be tested. Testing machines have needle probes which are pressed onto the bonding pads around the edge of the chip. There may be additional pads for test purposes that will not be connected externally. Not all the ICs on a wafer will work: those that fail the test are marked by the testing machine with a drop of ink for later rejection. The percentage of chips that pass the test is known as the *production yield*. In the early production batches of a new chip or process, the yield may be only a few % — nearly all the chips on a wafer are rejects.

At this stage in the fabrication process, the value of each IC chip is low: the most expensive part of the manufacturing process is yet to come. Up to this point hundreds of ICs have been processed together as part of a wafer and for many of the processing steps many wafers have been handled together. Each process operator therefore has been highly productive. The wafer is now scribed with a diamond scriber to separate the wafer into chips, each containing *one* integrated circuit. The wafer is broken into chips or *dice* by rolling a heavy steel roller weighing several kilograms back and forth across the wafer, which is protected by rubber sheets. From now on every chip is handled individually and costs are much higher. For this reason the wafers are likely to be air-freighted to a part of the world where labour rates are low, for this and subsequent stages of fabrication. An operator now sorts through the dice, selecting out the good ones from the rejects and picking them up with tweezers to stack them in a tray.

If the dice are to be sold unpackaged for use in hybrid microcircuits, they will be packed ready for shipment. Otherwise the dice will be mounted into an IC package. An electrical connection to the back of the chip is necessary for most types of IC and a common technique is eutectic bonding to a gold-plated header. Gold and silicon form a eutectic alloy at 370°C. The bonding pads on the chip now have to be connected to the external leads or pads of the package. This is done by hand, by an operator using micromanipulators working through an optical microscope. Gold wire is connected by ultrasonic welding or thermocompression bonding between the bonding pads on the IC and the package connections. The number of connections varies from 8 for a 741-type operational amplifier to 100 or more for a VLSI circuit. Once all the connections are in place, the package can be sealed, moulded or otherwise completed. The packaged ICs are then tested electrically again and may also be subjected to heat, vibration, pressure or vacuum to detect weak circuits which would otherwise fail early in life.

Hybrid microcircuits consist of a combination of bare monolithic IC chips with printed resistors and chip capacitors mounted on a ceramic substrate. Till and Luxon discuss the technology in detail.

Semiconductor Packaging

Figure 3.7 illustrates the variety of packaging styles used for ICs, while Table 3.1 summarizes the main characteristics of these package types. Hermetically sealed (airtight) metal cans were the earliest form of IC package and are still used for some linear circuits. Power voltage regulator ICs are often packaged in metal cans with a thick base similar to a power transistor package. The dual-in-line package (DIL) has been used since the 1960s for packaging both logic and linear circuits. It is designed for soldering into through-holes on a PCB or for insertion in a socket. Plastic DIL packages are moulded onto the metal lead frame after the chip has been bonded to the leads. Ceramic DIL packages are more reliable than plastic packages, but are also more expensive. They can be of either the frit-seal type, consisting of two ceramic slabs cemented together with a glassy ceramic onto a lead frame, or the side-brazed type with leads brazed onto connection pads along the sides of the package. The IC chip is mounted in a cavity in the ceramic and covered with a hermetically sealed metal lid in the case of a side-brazed package or with the top slab of ceramic in the case of the frit-seal package. The most common application for the frit-seal package is UV-erasable memories, where the chip is

See Fig. 6.1(b) for an illustration of a power transistor package.

Plastic packages can, for example, allow ingress of moisture along the leads, leading to corrosion of the leads or the IC bonding pads.

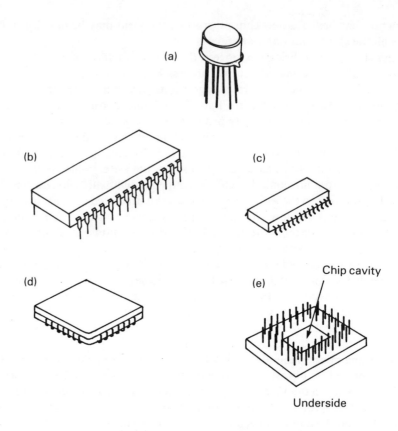

(a)

(b)

(c)

Chip cavity

(d)

(e)

Underside

Fig. 3.7 Integrated circuit packages: (a) hermetic metal can; (b) plastic DIL; (c) plastic small outline (surface mount); (d) PLCC; (e) PGA.

visible through a quartz window in the top slab.

When DIL packages were first manufactured ICs typically had 14 or 16 external connections, making a package about 20 mm × 7 mm. As LSI chips became available with 40 or more connections, the DIL package became unwieldy: a 40-pin DIL package measures about 50 mm × 16 mm, yet houses a chip about 5 mm square. To reduce the size of IC packages for VLSI chips, the pin-grid array (PGA) was developed, with pins still on a 2.54 mm (0.1 inch) pitch but arranged in several rows all around the underside of the package on a rectangular grid. PGAs are constructed in the same way as a side-brazed ceramic DIL package. The PGA can have up to 128 leads and yet be only 25 mm square.

At about the same time that pin-grid arrays were introduced, the technology of surface mounting was also being developed. Because several manufacturers were working on surface mounting simultaneously, two types of package have emerged. The small-outline (SO) package is essentially a scaled-down DIL package with leads spaced at half the pitch of a DIL package and folded out flat rather than projecting beneath the package. Most IC manufacturers have adopted the SO package for SSI and MSI logic and linear products. The chip-carrier package has leads spaced at the same 1.27 mm (0.05 inch) pitch as the SO package, but on all four edges. Plastic-leaded chip carriers (PLCCs) are moulded in a similar way to SO packages but have J-shaped leads rolled under the package body. Leadless ceramic

Table 3.1 Characteristics of integrated circuit packages

(a) Through-hole mounting types, 2.54 mm lead spacing

Type	Maximum number of leads	Features
Hermetic metal can	12	Leads on pitch circle
Plastic dual-in-line (DIL)	48	Low cost
Ceramic dual-in-line	64	High reliability
Ceramic pin-grid-array (PGA)	132	Low board area for number of pins

(b) Surface-mounting types, 1.27 mm lead or pad spacing

Type	Maximum number of leads or pads	Features
Plastic small-outline (SO)	28	Low cost
Plastic-leaded chip-carrier (PLCC)	124	Compact, low cost
Leadless ceramic chip-carrier (LCC)	124	High reliability

chip carriers (LCCs) have pads rather than leads and are similar to a side-brazed ceramic DIL package in construction. PLCCs and LCCs can be housed in sockets. SO packages are not socketable. Ceramic chip carriers are not suitable for soldering to epoxy PCBs because of the difference in thermal expansion coefficient of the ceramic relative to epoxy, which can cause stress-cracking of solder joints.

This point is discussed further in Chapter 9.

Handling of Semiconductor Devices

Many types of semiconductor devices and integrated circuits are damaged fairly easily by physical or thermal shock, overheating during soldering and, especially, by electrostatic discharge. The damage caused by mishandling is often not catastrophic – the device does not fail immediately, but is weakened by the damage, and eventually fails weeks or months later after being built into a product and shipped to an end user. Careful quality control is therefore essential on an electronics production line. Production operators must be made aware of handling precautions and provided with the correct tools and equipment for handling semiconductors. Completed electronic products can also be tested to reveal defects such as weakened semiconductor devices before they leave the factory.

Test methods are discussed in Chapter 9.

Thermal Damage

Germanium semiconductors were the earliest solid-state devices and were easily damaged by overheating during soldering. Although silicon devices are used for

Germanium diodes, for
example, find application in
some analogue circuits because
of their low forward voltage
drop compared to silicon
diodes.

nearly all applications today, a few germanium devices are still available and are used in specialized applications. Heat-absorbing pliers are recommended for holding the leads of these devices while soldering. Silicon semiconductor devices are more robust thermally but still require care if soldered by hand. Mass soldering is more easily controlled both in temperature and time, to keep within the limits specified in a data sheet (usually under the heading 'Absolute maximum ratings').

Electrostatic Damage

CMOS and MOS logic circuits
have gate-protection diodes
connected to external gate
electrodes to provide a path for
static charge to leak away. This
may not, however, prevent
damage caused by a discharge
into the gate terminal from an
external source.

Certain types of integrated circuits and discrete semiconductors are sensitive to damage by discharge of static electric charge. Not all device types are sensitive to damage (although none are completely immune). Field-effect devices with insulated gate electrodes are the most sensitive types. These include MOS and CMOS logic circuits and MOSFET transistors. Damage to the devices results from electrostatic discharge into the gate electrodes, causing dielectric breakdown and perforation of the gate insulation. Electrostatic charge can build up on clothing, shoes and floor coverings as a result of surfaces rubbing together. Figure 3.8 shows a typical special handling area (SHA) for electrostatic sensitive devices as recommended by the British Standards Institution. The workbench surface, compartment trays, floor mat and operator's stool are all electrically conductive. The operator is wearing cotton overalls and special shoes and is connected by a wrist strap to earth. Because of the danger of electric shock in an environment of earthed conductors, the electricity supply to the workbench is isolated from the mains supply by an isolation transformer and is fitted with a residual current device (RCD). The earthing straps have a resistance of 0.5–1 MΩ, sufficient to conduct static charge safely to earth, but high enough to limit current flow in the presence of

Electrical safety is discussed in
Chapter 10.

an electrical fault.

Electrostatically sensitive devices are protected in storage and in transit using special packaging materials. Dual-in-line ICs, for example, are inserted into conductive foam or kept in a conductive plastic tube. A fairly common method of protecting the leads of ICs is to insert them into a sheet of polystyrene foam: this must never be done with electrostatically sensitive ICs (and is perhaps best avoided totally in favour of conductive foam). Assembled PCBs containing sensitive devices can be placed in conductive plastic bags for shipment and storage.

Custom Integrated Circuits

There is a wide variety of standard integrated circuits available, both for logic and for analogue applications. If an electronic product is to be manufactured in quantity, however, it may be economically worthwhile to design a tailor-made *custom* IC specifically for the product. These circuits are known as ASICs, for application-specific integrated circuits. A custom-designed IC is cheaper per IC, more reliable and of smaller size and power consumption than the equivalent circuit implemented with standard SSI and MSI components. Also, the reduced size or greater functionality (or both) of the resulting product may give the company using a custom IC a market advantage over its competitors. There are several different approaches to custom IC design, each applicable over a certain range of production volumes, because production costs per IC are inversely related

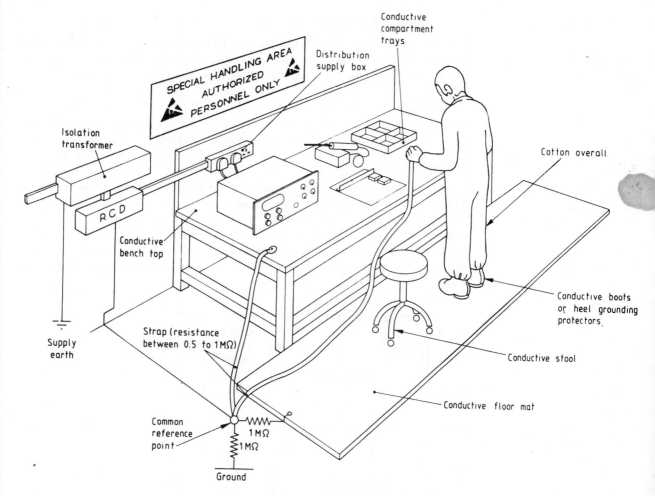

Fig. 3.8 A recommended special handling area for electrostatically sensitive devices. (From BS 5783:1984 with the permission of BSI)

to set-up and design costs. The most expensive option to set-up (full custom design) produces the lowest-cost ICs, but only if hundreds of thousands of ICs are to be made. Conversely, the cheaper techniques such as gate arrays are much less expensive to set up, but produce more expensive chips. For small production runs of perhaps 10 000 ICs, however, they may offer the cheapest total cost.

Full-custom Integrated Circuits

The most expensive form of ASIC is the full-custom integrated circuit. These are designed and manufactured in exactly the same way as standard ICs. They are justifiable only for very high-volume applications (hundreds of thousands of ICs per year) because of the high cost of design. They also are the cheapest type of custom IC, per unit manufactured.

Trade offs between *unit cost* and set-up or design costs and manufacturing quantity occur in all fields of engineering.

Companies using full-custom ICs in their products either have their own design and manufacturing facilities or contract out the work to a semiconductor manufacturer.

Standard-cell Integrated Circuits

Computer-aided design (CAD) has made possible a cheaper approach to custom IC design, known as the standard-cell system. A computer database holds a library of common circuit elements such as logic gates, flip-flops, operational amplifiers and so on stored in the form of their physical layout within the chip. The library is similar to the standard small- and medium-scale ICs used in non-custom design. A standard-cell library, however, can hold many more designs than are available as standard ICs. A standard-cell chip is designed by assembling together the required cells from the computerized library (using a CAD system). This involves not only the logical circuit design, but also the physical placement of cells on the chip. Since the shape and size of the cells is fixed there will be unused areas between cells. A standard-cell design will therefore occupy more chip area than a full-custom equivalent. Production costs will be higher than for a full-custom design because there will be fewer chips per wafer, but on the other hand design costs are lower because a lot of the design work has already been done in designing the standard cells (and the cost of *that* work is shared among all the purchasers of the cell library).

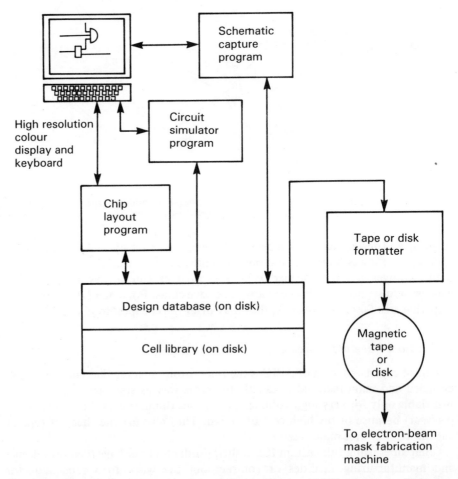

Fig. 3.9   Main components of a computer-aided design (CAD) system for IC design.

Fig. 3.10 A standard-cell chip. Note the unused areas and the different sizes of cells.
(Courtesy British Aerospace)

Once designed, a full set of masks has to be fabricated exactly as for a full-custom IC, and the manufacturing process is identical.

Gate Arrays

A third method of custom IC design and manufacture exists, with a lower design cost than the standard-cell system. It also differs from full-custom and standard-cell techniques in the manufacturing stages. The gate-array manufacturer designs a

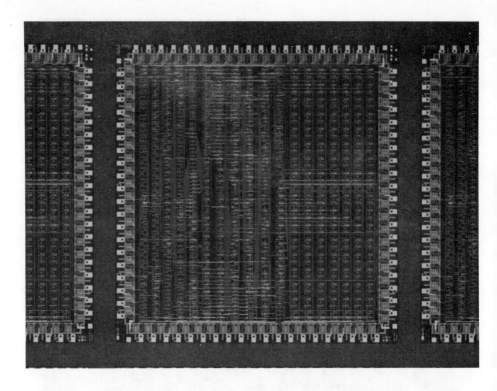

Fig. 3.11 A gate-array chip. Note the regular structure of array elements. This chip is fabricated in silicon-on-sapphire CMOS technology with two-level metallization. (Courtesy Marconi Electronic Devices Ltd)

standard chip with a fixed layout of logic gates and flip-flops and produces a full mask set with the exception of the metallization masks. Wafers are processed in quantity by the normal processes, but no metallization is applied. The customer designs the metal interconnect using CAD software supplied by the gate array manufacturer or else supplies the circuit diagram to the manufacturer who then designs the interconnect on the customer's behalf. A custom mask is then made for the interconnect and pre-processed wafers are metallized to the customer's design.

The setting-up costs of design and mask manufacture are comparatively low for gate arrays, so that a gate-array chip design may be viable for production quantities of only a few thousand per year. Gate arrays have several disadvantages, however, compared to standard-cell design, including longer interconnection paths on the chip due to the fixed layout of the array elements and the need to detail a complex design down to the elementary components making up the array.

Programmable Logic

The cheapest form of custom IC in terms of design costs is programmable logic. Several types of device exist, but all have in common the fact that they are manufactured in large quantities, fully packaged but uncustomized. They contain an array of logic gates interconnected by user-programmable links. The links can be fuses or insulated-gate MOS transistors which can be selectively burnt through or charged respectively to define the connectivity and therefore the function of the logic. The same technology can be used for programmable read-only memories (PROMs) used in computers and microprocessor systems to store machine-code programs and processor microcode.

Programmable logic and ROMs can also be fabricated as *mask-programmed* devices in which the logic function or memory contents are defined by the metallization mask, similar to the manufacture of gate arrays. Single-chip microprocessors with on-board ROM can also be mask-programmed. Mask-programming allows lower cost, high-volume production once the logic or program design has been proven.

Second Sourcing

Any electronics manufacturer, whether using standard ICs or custom-designed parts, needs more than one source of supply to guard against component shortages caused by technical or other problems at the IC manufacturer's facilities. The semiconductor industry recognizes this need and tries to set up *second sourcing* wherever possible. Typically, two semiconductor companies exchange mask sets so that each can manufacture and sell the other's designs. In some cases, very popular parts such as the 741 operational amplifier or the 555 timer IC, become available from five or more manufacturers. Problems can occur, however, especially with microprocessors and other LSI chips if the different manufacturers have designed their chips independently rather than exchanging mask sets, because of detail differences between the nominally identical products.

Summary

Integrated-circuit technology is a very important part of modern electronic engineering. It makes possible low-cost highly reliable products with a high level of functionality. Most integrated circuits are fabricated from silicon. Integrated transistors and other devices are formed within a silicon wafer by incorporation of dopants into the silicon crystal structure. The regions to be doped are defined by lithography and etching of a silicon dioxide layer grown on the surface of the wafer. Dopants can be introduced by diffusion or ion implantation. Additional silicon may be built up on a wafer by epitaxial growth, in which silicon atoms are deposited on the wafer from a gas. The epitaxial layer is an extension of the existing crystal structure of the wafer. Interconnections between devices on an IC may be made by a deposited metallization layer. Throughout all of these processes whole wafers containing hundreds of ICs are processed together often in batches of dozens of wafers. Once the wafers are broken up into individual dice after testing, each IC is handled individually for packaging and further testing, and manufacturing costs are high compared to the wafer processing stages. There are several different styles of IC package, each with their own advantages and disadvantages. These can be broadly divided into through-hole or surface-mounting types and into plastic and ceramic types. Surface-mounting types are smaller than equivalent through-hole types and ceramic packages are more reliable and expensive than plastic packages.

Semiconductor devices and ICs are sensitive to thermal and electrostatic damage and correct handling precautions must be taken in a production environment to reduce product reject rates.

ICs for custom applications may be designed in the same way as standard production ICs provided large quantities are required, or by assembling a design from pre-designed standard cells, or by designing a metallization layer to inter-connect a mass-produced, but unmetallized, gate array. Programmable logic can be regarded as a custom IC technology with lower setting-up costs than even gate-array technology, but with a higher cost per production IC.

Problems

3.1 At the time of publication of this book, VLSI ICs with 10^6 transistors per chip were reaching production. Using the smallest discrete transistors available, which are surface mount devices measuring about 2 mm × 3 mm, what area of PCB would be required to realize the equivalent function, making no allowance for interconnections?

3.2 If the trend illustrated in Fig. 3.1 were to continue until the year 2000 how many transistors could be expected on the largest chips then in production? Assume 10^6 transistors per chip in 1987 when this book was published and a doubling time of eighteen months.

Power Sources and Power Supplies

4

□ To introduce the main sources of electrical energy used in electronic systems, including mains supplies, batteries and photovoltaic cells.
□ To introduce the concept of a power supply.
□ To discuss the characterization and performance of power supplies.
□ To explain the functions of the main sub-circuits found in a power supply.
□ To explain the operation of linear and switching voltage regulators.

Objectives

All electronic circuits and systems require *energy* to operate. Energy is required to move electric charge, to produce heat, light or sound, to produce mechanical movement and to manipulate *information* (as in a computer). Energy is a *conserved* physical quantity: in a closed system energy can neither be created nor destroyed, although it can be converted from one form to another. In electronic engineering, we are usually concerned with electrical energy, although other forms of energy are also important. Heat, for example, is produced in electronic circuits, usually as a by-product of a useful function and is discussed in Chapter 6. Energy may be stored as chemical energy in a *cell* or *battery*. Chemical energy sources are discussed later in this chapter.

Energy is the capacity to do *work*. The SI unit of energy is the joule (J).

While the importance of energy should not be forgotten, electronics engineers more frequently use the concept of *power*. Power can be used to quantify the rate at which heat is produced in a resistor, the mechanical output of a motor or the rate at which an electronic system takes energy from its energy source.

Power is the rate of conversion, utilization or transport of energy. The SI unit of power is the watt (W) which is 1 J s^{-1}.

Let us examine the form in which electronic circuits of various types require their electrical energy supply. If we leave aside a.c. power control systems such as thyristor motor controllers, most circuits operate from one or more d.c. supply *rails*. Some circuits require only one rail and associated *return* (usually at 0 V), while others require two rails which are often symmetric with a 0 V return. Logic circuits and microprocessors, for example, usually operate from a single +5 V rail, while linear circuits such as active filters using op-amps require perhaps ±15 V rails. Most low-power electronic circuits use voltages below 20 V with respect to earth. Power amplifiers may require supply rails of up to 300 V. Cathode ray tubes (CRTs) used in televisions, oscilloscopes and visual display units may require voltages of up to 5 kV or more, although normally at low current.

The terms *a.c.* and *d.c.* stand for *alternating current* and *direct current* respectively. We customarily talk about a.c. voltage and d.c. voltage even though technically this is nonsense – what is an 'alternating-current voltage'? We can avoid the term *d.c.* by talking about *steady* voltages and currents, and in the frequency domain, *zero frequency*. For example: 'A low-pass filter has unity gain at zero frequency'.

Apart from a nominal d.c. voltage, what other characteristics of a d.c. supply rail need to be specified? Firstly, the voltage tolerance of the rail must be defined. A nominal +5 V rail supplying TTL logic, for example, could be allowed to vary from +4.75 to +5.25 V because this is the recommended operating voltage range for TTL circuits. Secondly, the allowable *ripple* on the rail must be defined. Ripple is a small a.c. component superimposed on the mean d.c. level and is usually due to the a.c. source from which the d.c. rail has been derived. Note that the ripple waveform is not usually sinusoidal. Ripple is specified by its frequency and its peak-to-peak amplitude.

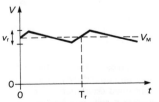

A typical ripple waveform with d.c. component V_M, ripple component v_r and ripple frequency $1/T_r$.

Energy Sources

As designers of electronic apparatus, we are concerned with the form in which energy is to be supplied to or contained within our equipment. The ultimate source of energy used by our design is of interest only in so far as this affects the form, availability and cost of the energy delivered to our system. We can distinguish two main forms of electrical energy which might be supplied to an electronic system: a.c. or d.c.; and several ways in which an energy source can be built-in to an electronic apparatus. Let us examine the characteristics of each of these.

A.c. Mains

Electronic equipment is often supplied with electrical energy from an external a.c. supply. The most common case is the mains supply derived from an electrical *grid*. The ultimate source of energy provided by a grid may be coal-fired, oil-fired, nuclear or hydroelectric power stations, diesel or gas-turbine generators or wind turbines. Grid supplies are characterized (in the developed countries at least) by high reliability and low-cost energy (compared to other sources). A.c. supplies may also be found on board trains, aircraft and ships where (except in the case of electric trains) the supply will be provided by generators driven from the engines.

A.c. supplies are widely used because of the efficiency and ease with which a.c. can be *transformed* from one voltage to another. Alternating supplies are characterized by their voltage and their frequency. Three common nominal frequencies are used: 50 Hz on the grid systems of the UK, Europe and Australia, 60 Hz on the North American continent and 400 Hz on board aircraft. The higher frequency used on aircraft is due to the reduced size and weight of transformers operating at the higher frequency.

Nominal mains voltages vary from one country to another. 240 V a.c. r.m.s. is normal in the UK, although 110 V is often found on building sites and in some factories where the lower voltage is used for safety reasons. In North America 115 V a.c. r.m.s. is commonly used. A.c. supplies may vary in voltage, frequency, or both, from their nominal values. These variations are due to *load variations* on the supply – it is common, for example, for the UK grid frequency to drop by up to 1.5 Hz when a very popular television program ends. This effect is caused by millions of people switching on electric kettles to make cups of tea! The frequency drop is usually corrected within a few seconds as the power stations supply more power. Local voltages drops may also occur, particularly in premises located some distance along a supply cable, as consumers nearer the substation switch on heavy loads. On small electricity grids voltage and frequency variations may be more marked, and the design of electronic equipment may have to take this into account.

Apart from voltage and frequency fluctuations, a.c. supplies may also be subject to momentary interruption (as, for example, when the local electricity board switches a substation from one circuit to another) and may also carry *electrical interference* in the form of high-frequency periodic disturbances, switching transients, or harmonics of the grid frequency. This interference may be caused by other electrical or electronic equipment connected to the supply or by natural phenomena such as lightning discharges, and could have a serious effect on the operation of electronic systems which are not designed to cope with it.

Electronic equipment is usually operated from a *single-phase* a.c. supply.

Electrical safety is discussed in Chapter 10.

R.m.s. stands for *root mean square* – the square root of the time-averaged value of the square of a waveform:

$$V_{rms} = \sqrt{(\bar{V}^2)}$$

For a sinusoid, the r.m.s. value is $1/\sqrt{2}$ times the peak value, so the peak value of the UK 240 V a.c. r.m.s. mains supply is $240\sqrt{2}$ V or 339 V.

Harmonics of the grid frequency may be caused by thyristor-controlled equipment. For further details see Bradley.

Internal Energy Sources

If a piece of electronic equipment has to be self-contained, with its own power source, there are only two options (neglecting portable diesel engines and generators). These are electrochemical cells and photovoltaic cells. A hand-held VHF transceiver, as used by police forces, is an example of a self-contained equipment powered by electrochemical cells. *Electrochemical cells* directly convert chemical energy into electrical energy. They can be divided into two types: batteries and fuel cells. *Photovoltaic cells* directly convert the energy of visible or ultraviolet light into electrical energy. They are, of course, tremendously important on space-craft such as communications satellites.

Fuel cells convert chemical energy from reactants supplied externally to the cell into electrical energy. Their best known application is on board manned space-craft, such as the American Apollo and shuttle craft where hydrogen–oxygen fuel cells supply electrical power (and drinking water as a useful by-product). Fuel cells will not be covered further in this book.

Batteries

Batteries are closed electrochemical power sources. They convert chemical energy from reactants incorporated into the device during manufacture to electrical energy. Originally the term 'cell' was used in this context, a battery being a collection of cells wired in series or parallel. In modern usage, cells are often referred to as batteries.

Two main types of battery exist, known as *primary* and *secondary*. Primary batteries can be used once only: when the chemical reactants have been consumed as a result of electrical discharge of the battery, the device must be discarded. Secondary batteries are based on a reversible chemical reaction: the battery may be recharged by passing electrical current through the device in the opposite direction to the discharge current. Despite their ability to be recharged, secondary batteries have a finite useful life: eventually recharging fails to store sufficient chemical energy in the battery for useful operation.

Different types of battery have different *nominal open-circuit voltages*, depending on the electrochemical reaction used in the cells and the number of cells in the battery. The actual voltage provided by a battery falls as the energy stored in the battery is used. The change in voltage is shown on a *discharge characteristic* for stated conditions of discharge such as continuous discharge into a specified resistance or intermittent discharge for a specified period per day.

When comparing and selecting batteries we are usually interested in the amount of energy that the battery can supply before it is fully discharged. This quantity is called the *capacity* of a battery and can be stated in watt-hours (W h), which is a unit of energy, or more frequently, in ampere-hours (A h), which is a unit of electric charge.

Since a battery normally has a fairly constant voltage during discharge, we can calculate the approximate energy content of a battery by multiplying its capacity in A h by its nominal voltage, remembering that an ampere-hour is 3600 coulombs because there are 3600 seconds in an hour.

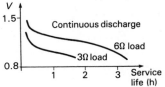

A typical discharge characteristic for two different loads.

Whether the capacity is stated in Wh or Ah, it is a measure of the energy available from the battery, not its ability to store electric charge (as in a capacitor): the mechanism of energy storage in a battery is chemical, not electrical.

Worked Example 4.1   Calculate the energy content of a 2 A h 12 V battery in (a) watt-hours and (b) joules.

Solution
(a) 2 A h × 12 V = 24 W h
(b) 2 A h × 3600 × 12 V = 86.4 kJ.

As in many other areas of electronic engineering, the concept of a normalized quantity is useful when comparing different systems. Battery discharge rates are often expressed in a normalized form known as the 'C' rate. A discharge current of 1 C will discharge a battery in 1 hour. A rate of C/5 will discharge a battery in 5 hours and a rate of 5 C will discharge a battery in 12 minutes. For a 2 A h battery, the C/5 rate is 400 mA. The C rate can also be used to quantify charge rates for secondary batteries.

> The concept of the C rate is an approximation, valid only over a limited range of discharge rates.

The capacity of a battery is not a precise measure of the energy available because the amount of energy that can be extracted from the battery depends upon how it is used. Some types of battery provide more energy in total if they are used intermittently and are allowed to 'rest' between discharges, while others are more suited to continuous discharge at a steady rate. When discussing capacity it is usual to say that the capacity of a battery is dependent on the pattern of discharge, although in reality, it is the amount of energy that can be extracted that is variable.

All batteries will self-discharge to some extent when not in use, because chemical reactions occur within the battery even when no current is being drawn. This limits the performance of secondary batteries and the shelf-life of primary batteries (typically to a few months or years).

Worked Example 4.2   A portable VHF radio transceiver consumes 3 W of power at 15 V when transmitting and 0.5 W when receiving only. If the unit is to operate from a secondary battery over an 8-hour shift, what battery capacity is required, assuming the radio is transmitting for a total of 30 minutes during the shift? What has been assumed about the battery capacity?

Solution
Transmitting:

$$\frac{0.5 \text{ hours} \times 3 \text{ W}}{15 \text{ V}} = 0.1 \text{ A h}$$

Receiving only:

$$\frac{7.5 \text{ hours} \times 0.5 \text{ W}}{15 \text{ V}} = 0.25 \text{ A h}$$
$$\text{Total energy} = 0.35 \text{ A h}$$

A battery of 0.35 A h capacity is needed, assuming that the intermittent loading will not reduce the battery's capacity and making no allowance for loss of capacity with life.

Main Primary Battery Types

The most widely used primary battery is the Leclanché cell and its variants. Originally, Leclanché cells were wet systems in glass jars, but nowadays the cells are made using chemical reactants in paste form and are referred to as 'dry' cells. There are three cell types in common used based on the Leclanché cell. The first of these is known as a *zinc-carbon* cell from its construction with a carbon rod electrode down the centre of a zinc can which serves as a case and the outer electrode. The paste electrolyte in the cell consists of ammonium chloride and zinc chloride mixed with manganese dioxide. This is the cheapest of the three Leclanché cells and is best suited to applications where current is drawn from the battery intermittently (say for 1 hour per day). The available capacity of the zinc-carbon cell varies with the pattern of discharge, so that it is not possible to state a single capacity for a cell. The capacity is also reduced at low temperatures, making these cells unsuitable for use in equipment to be used out of doors in freezing conditions. The second variant of the Leclanché cell is known as *zinc chloride* and has all the ammonium chloride of the zinc-carbon cell replaced by zinc chloride. This gives greater capacity at high current drains and better low-temperature performance, although at greater cost than the zinc-carbon cell. The third variation on the Leclanché cell is the *alkaline-manganese* cell which has a potassium hydroxide (alkaline) electrolyte. These cells are suitable for continuous high-current discharge and have about four times the capacity of a similar sized zinc-carbon cell used in this way. Their low-temperature performance is similar to that of zinc

Table 4.1 Main primary battery types

Type	Nominal open-circuit voltage	Main characteristics
Zinc-carbon (Leclanché)	1.5 V	Low cost. Best used intermittently. Poor performance at low temperature.
Zinc chloride (Leclanché)	1.5 V	Improved capacity at high current drain compared to zinc-carbon and better low-temperature performance.
Alkaline-manganese	1.5 V	Suited to high-current continuous discharge. Long shelf life. Similar low-temperature performance to zinc chloride but at higher cost.
Zinc-mercuric oxide	1.35 V	Available as 'button' cells for miniature equipment such as watches. Flat discharge characteristic. Good shelf life.
Zinc-silver oxide	1.6 V	High cost. Flat discharge characteristic. High capacity per unit mass in large sizes. Military applications.

chloride cells but at higher cost. The shelf life of alkaline–manganese cells is good — they will retain up to 80% of their initial capacity after 4 years in storage at 20°C. All three Leclanché types have a nominal open-circuit voltage of 1.5 V.

For miniature equipment such as watches and calculators there are two primary cell types manufactured as 'button' cells. These are the *zinc–mercuric oxide* cells with an open-circuit voltage of about 1.35 V and the *zinc–silver oxide* cell with an open-circuit voltage of about 1.6 V. Both types have a flat discharge characteristic: the cell voltage remains fairly constant until the battery is almost fully discharged. Zinc–silver oxide batteries are also made in large sizes for military applications such as missiles and torpedoes, where they are used for their high capacity per unit mass despite their high cost.

Table 4.1 summarizes the main types of primary battery.

Main Secondary Battery Types

The two main types of secondary battery are the lead–acid type used in road vehicles for starting, lighting and ignition (SLI) and the nickel–cadmium type used in aircraft and military vehicles. Both are also available in smaller sizes for powering portable equipment.

The lead–acid cell consists of metallic lead electrodes and sulphuric acid electrolyte. Lead–acid cells have a nominal open-circuit voltage of 2 V. Over 30% of world lead production is used in batteries, about 100 million units per annum being produced. The capacity of lead–acid batteries drops very rapidly below 0°C. Vehicle batteries account for over 80% of lead–acid battery production. They must support short intense discharge of up to 5 C on engine starting. Car batteries are rated at 30–100 A h at 12 V, while commercial vehicle batteries have capacities of up to 600 A h at 24 V. Larger batteries are used for traction applications such as milk-floats and railway locomotives.

For portable equipment, sealed lead–acid batteries are available with capacities from 2 to 30 A h. These are of lower cost than nickel–cadmium batteries, but are heavier for the same capacity. They have a life in excess of 300 charge–discharge cycles and are maintenance free, needing no topping-up of the acid electrolyte. They exhibit a fairly constant voltage during discharge at up to the C/4 rate, and can withstand short high-rate discharge.

Nickel–cadmium batteries are based on cadmium and nickel oxide electrodes with a potassium hydroxide electrolyte. The open-circuit voltage is about 1.2 V. Nickel–cadmium batteries are more expensive than lead–acid batteries and are the most important alkaline secondary type, accounting for over 80% of all alkaline secondary batteries sold in the West. Unlike lead–acid cells, they can work well at temperatures down to less than −30°C. They have a flat discharge characteristic and can accept continuous overcharging at a low charge current. (This is known as 'trickle' charging). Small nickel–cadmium batteries of three cells with a nominal voltage of 3.6 V are used on printed-circuit boards to supply back-up power to CMOS memories. They are trickle charged from the memory supply rail when mains power is available and ensure data retention in the memory when it is not.

For further reading on batteries see Vincent.

Photovoltaic Cells

Photovoltaic cells convert the energy of visible or ultraviolet light into electrical energy. They are used on board communications satellites to supply electrical power where no other energy source could supply the power needed over the many years of the satellite's life. Solar cells may also be used on Earth, for example to power communications relay stations located in remote regions far from electrical grids. In this case electrochemical cells are needed to store energy for use during the night when direct solar power is not available. On a smaller scale, some electronic watches and calculators are powered by photovoltaic cells with secondary electrochemical cells providing energy storage to power the device while in darkness or low light.

The most important photovoltaic cells are silicon junction diodes of large area with a thin n-type region on the exposed face.

A p–n junction with no externally applied bias develops a space-charge region on either side of the junction as mobile charge carriers diffuse across the junction under the influence of concentration gradients. In equilibrium, there is no net charge transport across the junction because of the presence of an electrostatic potential.

If now, electromagnetic radiation of suitable wavelength illuminates the junction, electron–hole pairs can be created by photon absorption. This process can be thought of as the ionization of a silicon atom, creating a free electron and a positively charged silicon ion. The silicon ion can attract an electron from a neighbouring atom, so that the positive charge (a hole) is mobile. The free electron and the free hole are, however, influenced by the electrostatic potential difference across the junction: they move in opposite directions, the electron towards the cathode and the hole towards the anode. If the diode is connected to an external circuit, current can flow and supply power to the external circuit. The direction of this current flow is that of a *reverse* current through the diode: the positive potential is developed at the *anode*. This means that in a circuit where a photovoltaic cell charges a secondary battery a blocking diode must be connected in series with the cell to prevent the battery from driving *forward* current through the cell during darkness. The blocking diode must have a low forward-voltage drop and often a Schottky barrier (semiconductor–metal) diode is used. The e.m.f. generated by a silicon photovoltaic cell is around 0.5 V, so that practical circuits using secondary batteries have several photovoltaic cells in series to raise the e.m.f. to a practical level.

Power Supplies

So far in this chapter, we have looked at energy sources and the forms in which energy may be used by electronic circuits. We can now look at the way in which electrical energy can be converted into the required form. In many electronic systems this conversion is performed by a sub-system called a *power supply*. Even in battery-operated equipment where the energy source provides energy in almost the required form, there may still be some form of power supply.

A wide range of power supplies are available commercially, usually from manufacturers specializing in this field. Many larger electronic systems use commercial power supply units bought off the shelf. In other cases, however, a custom design may be needed, because for example, a special physical form, low weight or high reliability is required. For high-volume mass production, a custom power supply, as with any other component or sub-system, may be cheaper than any commercial

Other cells are based on cadmium sulphide, selenium and gallium arsenide. All include a p–n or semiconductor–metal junction.

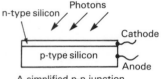

A simplified p–n junction photovoltaic cell.

There are other, equivalent, explanations of p–n junction operation in terms of electric field. Any book on electronic or semiconductor physics will discuss these.

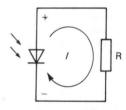

The direction of current flow from a photovoltaic cell.

The arrow in the diode (and junction transistor) symbol represents the direction of forward current flow when the diode is driven by an external circuit. Conventional current, by historical accident, flows in the opposite direction to electrons.

The abbreviation PSU for 'power-supply unit' is often used.

off-the-shelf design because of a better match between the power-supply design and the system requirement. Often such custom design is contracted out to a power-supply firm with special expertise.

We can now look at the main functions of a power supply. The most obvious and common function of a power supply is to convert electrical energy at the source voltage to some other voltage, higher or lower than the source voltage with or without a change from a.c. to d.c. or *vice versa*. A microcomputer, for example, might be designed to operate from an a.c. mains supply and yet contain circuitry operating at 5 V d.c. A power supply would be needed therefore to reduce the voltage and convert the energy to d.c. Within the same microcomputer, there might also be a cathode-ray tube (CRT) display requiring a low-current but high-voltage supply (say 4 kV) for the electron gun, requiring an increase in voltage and conversion to d.c. On board a communications satellite, d.c. will be available from batteries charged from a solar panel. For economy of space and weight, only one voltage will be available from the batteries. Electronic circuits and systems such as microwave amplifiers, attitude controllers and computers, however, may need a variety of voltages both lower and higher than the battery voltage. *D.c.–d.c. converters* can produce these voltages at high efficiency without wasting valuable energy as useless heat. *Inverters* generate a.c. from a d.c. input. One common application is on board caravans and boats to generate 240 V a.c. 50 Hz from a 12 V battery, allowing low-powered domestic mains-operated equipment such as radios and shavers to be used.

Voltage conversion, whether to a higher or lower voltage, is possible in practical terms only in an a.c. circuit, using a transformer. D.c.–d.c. power supplies (or converters) are in reality, then, d.c. to a.c. to d.c. power supplies.

Power supplies operating from an a.c. source also have to provide energy storage during the parts of the source cycle where little or no energy is available from the supply. It may also be necessary to store energy to supply the output current during momentary loss of the a.c. supply, such as happens when a substation is switched. Energy storage is usually in the form of electric charge in the power supply's reservoir capacitors and is usually practical for only a limited time in most systems. Longer-term energy storage requires the use of batteries. A power supply with batteries is an example of an uninterruptible power supply (UPS). A UPS produces constant output even during breaks in the a.c. mains supply.

A change in voltage is always accompanied by a compensating change in current: the power output of a power supply is always less than the power input.

Voltage increase at low current can be achieved without a transformer using a diode multiplier such as the Cockcroft–Walton multiplier. These circuits do, however, require a switching action.

Larger uninterruptible power supplies may use diesel generators as well as, or instead of, batteries.

Exercise 4.1 A microcomputer uses 4 W of power at 5 V d.c., derived from a mains supply. The microcomputer is to be capable of continued operation despite momentary interruptions in the mains supply lasting up to 125 ms. Show that a capacitor of about 1 F is needed if connected across the 5 V rail, assuming that the rail voltage must not drop below 4.9 V during the supply interruption.

Examples of loads which would demand varying power are: a power amplifier driving a loudspeaker with speech; TTL logic circuits, where the power demanded depends on the numbers of logic gates in the '0' and '1' states.

The third main function of a power supply is maintenance of a constant output irrespective of varying *load*. This is known as *regulation* or *stabilization* and a power supply having this feature is called a regulated or stabilized power supply. Not all power supplies have this feature as some applications can tolerate variations in voltage or current. A d.c. supply for low-voltage filament lamps and relays is an example of such an application.

Power-supply performance can be represented graphically as a *characteristic*,

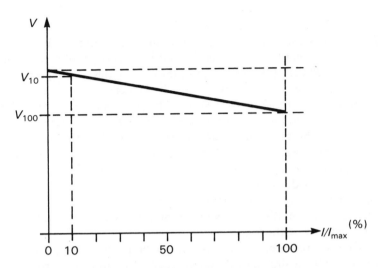

Fig. 4.1 Output characteristic of a voltage-regulated power supply showing quantities used in defining load regulation.

which shows output voltage as a function of output current. Ideally, the output voltage would be independent of output current, but in practice some drop in voltage is unavoidable and is quantified by a parameter known as *load regulation*. The definition of load regulation varies somewhat but a fairly common definition is given below and in Fig. 4.1. V_{10} is the output voltage at 10% of the full load current and V_{100} is the output voltage at 100% of the full load current. Load regulation is then defined as

Varying definitions of such practical parameters are quite common – never be tempted to assume what someone means by such a term!

$$\text{Load regulation} = \frac{(V_{10} - V_{100})}{V_{10}} \times 100\% \qquad (4.1)$$

Load regulation is normally expressed as a percentage. An alternative definition might use the full load and no-load voltages, rather than the full load and 10% of full load voltages.

Output voltage variation might also be due to variation in the source voltage. The a.c. mains voltage, for example, can vary, particularly for consumers at the end of a cable some distance from a substation. The terminal voltage of a battery falls as the battery becomes discharged. It is possible to define a regulation factor to quantify the performance of a power supply under variations of input voltage, but this can be done in many ways and it will not be discussed further here.

A final factor which can influence the regulation of a power supply is ambient temperature. Normally, the variation in output voltage with temperature is small and roughly linear over the operating temperature range of the power supply, so that the normal idea of a temperature coefficient can be used to quantify this aspect of the power supply's performance.

Apart from the precisely controlled voltage, there is another benefit to be had from a voltage-regulated power supply. This is a low a.c. or dynamic impedance and is very important in the operation of linear circuits. In analysing, say, a common-emitter amplifier circuit we assume the power supply to be of negligible impedance in deriving our small-signal model. We must not forget that this assumption may not always be valid.

Overload Protection

We can now turn our attention to the part of the output characteristic beyond the full-load current. What happens when the full-rated current of a regulated power supply is exceeded? Fig. 4.2 shows some of the possibilities. If no special provision is made in the design of a power supply, damage or destruction of some components may occur due to overheating. The power supply may include a *fuse* which will rupture and disconnect the power supply from its energy source. This possibility is not shown in the figure. *Current limiting* is the next most likely possibility — the power supply will be designed so that beyond full-rated load current, the output voltage will be reduced. This may be done by limiting current as in Fig. 4.2(b). A third, more elaborate possibility is known as *foldback limiting* and is illustrated in Fig. 4.2(c). Here, once the full rated load current has been exceeded, the power-supply output voltage and current are reduced, bringing both down independently of the load.

We have considered protection of the power supply itself from damage due to excessive current being drawn by the load. Foldback limiting also protects the load

Current limiting as shown in Fig. 4.2(b) is a popular option for bench power supplies. Often the limiting current is adjustable so that limiting can be used to protect the circuit being tested.

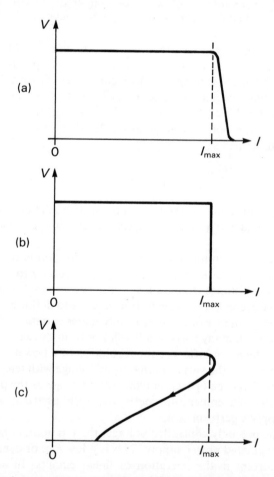

Fig. 4.2 Output characteristics of voltage-regulated power supplies: (a) with voltage reduction beyond full-load current; (b) with cross-over to constant current at full-load current; (c) with 'fold-back' limiting.

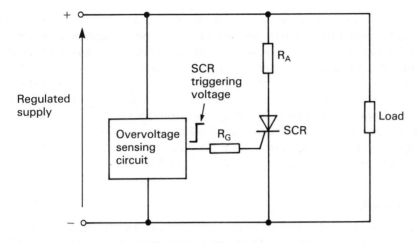

Fig. 4.3   A crowbar overvoltage protection circuit.

to some extent. One other eventuality that must be considered, however, is *over-voltage*. Suppose that a regulated power supply supplies current to a large system containing many expensive integrated circuits (such as a computer). If the power supply becomes faulty and the output voltage becomes too high, hundreds or thousands of pounds worth of circuitry could be damaged beyond repair if the voltage exceeded the absolute maximum ratings of the integrated circuits. To guard against this prospect, *overvoltage protection* is normally included in the power-supply design. Figure 4.3 shows a common solution known as a *crowbar circuit*. The essence of this circuit is to create a short circuit (or at least a very low-resistance path) across the power-supply terminals as soon as an overvoltage is detected (within microseconds). When the overvoltage sensing circuit detects that the regulated voltage has risen beyond its normal limit a triggering signal is generated, switching the SCR or thyristor into its conducting state. R_A limits the maximum current through the SCR and is typically a fraction of an ohm. Once triggered, the SCR effectively short circuits the regulated supply and prevents damage to the load. The SCR continues to conduct until current is switched off.

Overvoltages may also occur at switch-on and switch-off unless the power-supply design deliberately includes proper control such as a defined 'power-up' sequence.

The characteristics of thyristors or SCRs are discussed by Bradley.

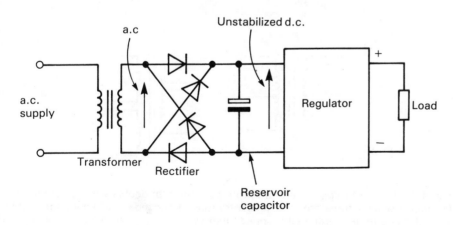

Fig. 4.4   Block arrangement of a typical mains-operated d.c. power supply.

We have now examined the main functions and characteristics of power supplies and can turn our attention to the circuits used within power supplies. Figure 4.4 is a block diagram of a typical mains-operated power supply. The transformer changes the input a.c. voltage to a higher or lower a.c. voltage, which is then converted to d.c. by the full-wave bridge rectifier and reservoir. At this stage, before the regulator, the d.c. voltage is unstabilized (it will vary as the load current varies) and may also have greater ripple than the regulated voltage. The regulator stabilizes the load voltage and may also include overload protection circuits. Practical power supply designs may not show such a clear distinction between the functional blocks. Let us now consider each functional block in turn.

Transformers

See for example Senturia and Wedlock.

Transformers are widely available commercially in a range of physical sizes, power ratings and electrical configurations and are designed for operation at a specific frequency (usually 50, 60 or 400 Hz). The theory of transformer operation is covered elsewhere, and will only be discussed briefly here. Transformers consist of one or more electrical *windings* of low d.c. resistance, wound onto a *core* of magnetic material (commonly iron). A changing current in one winding *induces* a changing magnetic field in the core which links the same or another winding and induces a changing e.m.f. in that winding. Two main types of transformer are shown in Fig. 4.5. The *autotransformer* (Fig. 4.5(a)) has one winding with intermediate tappings, while the *double-wound* transformer has separate *primary* and *secondary* windings. The double-wound transformer is almost universally used in

The primary winding is the energy input side of the transformer. The secondary winding is the energy output side of the transformer.

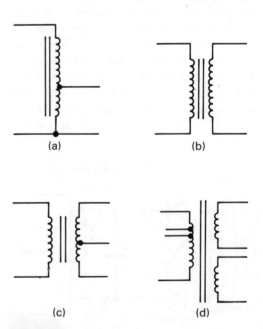

Fig. 4.5 Types of transformer: (a) autotransformer; (b) double-wound transformer; (c) double-wound transformer with centre-tapped secondary winding; (d) double-wound transformer with multi-tapped primary winding and separate secondary windings.

electronic power supplies, because of the increased safety obtained by having electrical isolation between the windings. The ratio of turns on the secondary winding to the number of turns on the primary winding determines the ratio of secondary voltage to primary voltage. The amount of power that can be drawn from a secondary winding is always less than the power supplied to the primary winding as some energy is lost as heat in the windings due to their resistance (copper loss), and as heat due to circulating electric currents (eddy currents) in the core (iron loss). Iron losses can be reduced by constructing the core from thin flat sheets of iron, or laminations, stacked together and insulated from each other with a layer of varnish or by constructing the core from *ferrite* which is non-conductive. Copper losses can be reduced by increasing the cross-sectional area of the wire used for the windings. Copper is an expensive metal, however, so the transformer designer must trade off copper losses against the cost of the transformer. For small transformers such as are used in electronics, cost is likely to be more important than a small energy loss.

Ferrites are discussed in Chapter 5.

There are three common constructional designs for double-wound transformers shown diagrammatically in the margin. The shell type is the most common for small transformers and has all the windings wound on a common centre *limb* of the core. Small core-type transformers are less common and have primary and secondary windings located on separate limbs of the core. The chief advantage of this arrangement is reduced capacitive coupling between the windings. The third transformer shown consists of a toroidal (doughnut-shaped) core and is a compact design popular for low-profile equipment. It also has reduced flux leakage compared to other designs, a factor which can be important in audio-amplifiers and low-frequency instruments because of their tendency to pick up and amplify mains frequency signals. When mounting toroidal transformers it is most important not to create an electric circuit through the centre of the toroid. This would constitute a short-circuited secondary and would cause overheating and possibly destruction of the transformer. Mountings should either use non-conducting fasteners (such as nylon screws) or be electrically insulated at one or both ends.

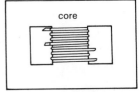

'Shell' type transformer with all windings on centre-limb.

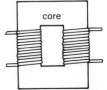

'Core' type transformer with primary and secondary windings on separate limbs.

Toroidal transformer.

The isolation between the primary and secondary windings of a transformer can be less than perfect at higher frequencies because of capacitive coupling between the primary and secondary windings. This coupling may result in transmission of electrical interference (unwanted noise and transients) into the secondary circuit. To reduce the coupling effect, some transformers are fitted with an electrostatic screen between the primary and secondary windings, brought out to an external terminal which is usually grounded.

Transformer ratings are usually expressed in VA (volt-amperes), rather than watts, or by stating the maximum r.m.s. current rating of each winding. Transformer secondaries should be fused, or otherwise protected against damage by a short-circuit load because a secondary winding can be supplying heavy current under fault conditions without the primary current exceeding the rating of the primary winding. It is therefore not sufficient to rely on a fuse in the primary circuit to protect the secondary.

The number of volt-amperes (VAs) is the product of the r.m.s. current and the r.m.s. voltage. It is not equal to power (in watts) because the current and voltage are out of phase in an a.c. circuit (unless the circuit is purely resistive).

Rectification

Rectification is the conversion of a.c. to pulsed d.c. The most important rectifying component in modern use is the semiconductor diode. Power diodes may

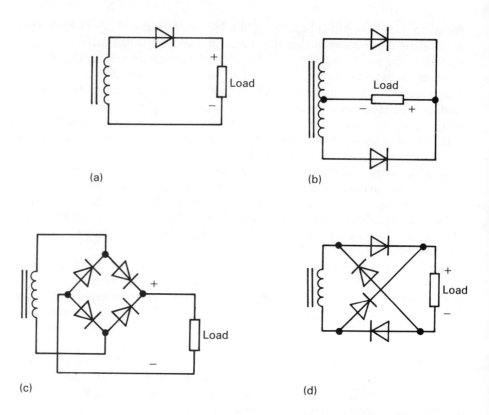

Fig. 4.6 Rectifier circuits: (a) half-wave; (b) full-wave; (c) full-wave bridge; (d) alternative representation of the full-wave bridge.

sometimes be referred to as rectifiers to distinguish them from signal diodes. Full-wave rectification is universally used in electronics and produces pulsed d.c. at twice the supply frequency. This can be achieved with two diodes if a centre-tapped transformer is used, or with four diodes connected in a *bridge* configuration. The bridge configuration is the most widely used nowadays, requiring a secondary winding without a centre tap on the transformer. Four diodes arranged in a bridge with four terminals or leads are commonly available and known as *bridge rectifiers*.

Rectifier or power diodes have separate ratings for average forward current and surge current. As will be seen in the next section, the inclusion of reservoir capacitors in power-supply circuits can cause large surge currents to flow when the diodes start to conduct in each cycle. Power diodes also have greater forward voltage drop than small-signal diodes so that power dissipation in a rectifier diode can be significant. Lastly, the maximum reverse voltage or peak inverse voltage (PIV) rating of rectifiers must be adequate both for the normal reverse voltages present in the circuit and for any abnormal transient voltages on the supply.

Reservoir Capacitors

For the majority of electronic applications, the pulsed d.c. output from a rectifier is unsuitable directly. In some applications a *reservoir capacitor* or smoothing

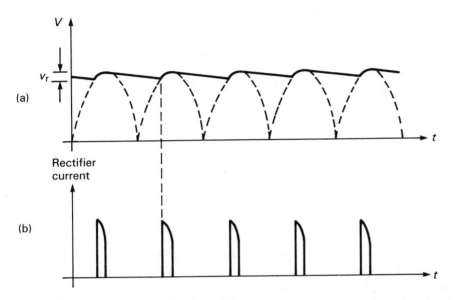

<figure></figure>

Fig. 4.7   (a) Output waveform of a full-wave rectifier circuit with reservoir capacitor;
(b) rectifier current.

capacitor connected between the supply rails is sufficient.

Figure 4.7 shows the effect of a reservoir capacitor on the output of a full-wave rectifier. The *ripple* voltage v_r is determined by the value of the reservoir capacitor, the load current and the supply frequency (which determines the time over which the capacitor discharges).

Worked Example 4.3

A reservoir capacitor is to be connected across the output of a full-wave bridge rectifier. Calculate:

(a) the phase angle at which the rectifiers start to conduct;
(b) the value of capacitor required;
(c) the peak rectifier current, neglecting any series resistance in the circuit.

The peak value of the rectified waveform is 6.5 V, the supply frequency is 50 Hz. The ripple must not exceed 50 mV peak-to-peak and the load current is 100 mA maximum.

Solution

(a) The phase angle is $\sin^{-1}(6.45/6.5) \simeq 83°$.
(b) Assume the capacitor discharges linearly between the peak of one cycle and the angle in (a):

$$C = \frac{\Delta q}{\Delta V} \simeq \frac{100\ \text{mA} \times 10\ \text{ms}}{50\ \text{mV}} \simeq 20\ \text{mF}$$

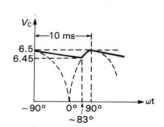

(c) The capacitor voltage V_C is given by

$$V_C = 6.5 \sin \omega t, \text{ where } \omega = 2\pi \times 50 \text{ rad/s} \quad 83° \leqslant \omega t \leqslant 90°.$$

The rate of change of V_C is $\dfrac{dV_C}{dt} = 6.5\omega \cos \omega t$ and the greatest rate of change occurs when $\omega t = 83°$.

$$\left.\frac{dV_C}{dt}\right|_{max} = 6.5(2\pi) \times 50 \times \cos 83° \simeq 250 \text{ V s}^{-1}$$

The peak rectifier current is given by:

$$I_R|_{max} = C \frac{dV_C}{dt} + 100 \text{ mA} = 20 \text{ mF} \times 250 \text{ V s}^{-1} + 100 \text{ mA}$$

$$\simeq 5.1 \text{ A}$$

There are low-voltage reference diodes (which are actually integrated circuits) which use the bandgap energy of silicon to provide a reference voltage. These are known as bandgap references.

All junction diodes exhibit reverse breakdown, but usually at much higher voltages. Zener diodes are specially fabricated to break down at a precise voltage and to withstand continuous power dissipation in the breakdown region. The name *Zener diode* is actually a misnomer for diodes with breakdown voltages above about 5 V. The Zener effect occurs when electrons within the space-charge region of the diode are dislodged from their atoms by the electric field intensity. The dominant process in most diode breakdown is *avalanche multiplication*, where the energy of dislodged electrons is sufficient to dislodge further electrons from their bonds.

Voltage References

All regulated power supplies require a voltage reference: a device or circuit which can maintain a constant voltage between its terminals independently or nearly independently of variations in current or ambient temperature. Before considering the various types of regulator, let us consider the most commonly used voltage reference component, the Zener diode.

Figure 4.8 shows the *V–I* characteristic of a typical Zener diode. The characteristic shows only the reverse-biased region because Zener diodes are operated in reverse breakdown. Their forward characteristic is of no interest. The main feature of the characteristic to note is that, for voltages greater than the breakdown voltage, V_Z, a small increase in voltage produces a large increase in current. Looked at another way, the voltage across the diode is nominally constant over a wide range of currents.

The *V–I* characteristic in the breakdown region is not precisely parallel to the

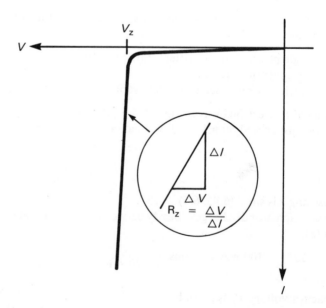

Fig. 4.8   *V–I* characteristic of a Zener voltage-reference diode.

current axis (a change in current *does* produce a small change in voltage). This variation is expressed as the diode's *slope resistance* R_Z (in ohms) and is the ratio of incremental voltage change to incremental current change at a specified current. R_Z is not constant, but varies with current. Typically, Zener diodes have slope resistances from a few ohms to a few tens of ohms. Breakdown voltages also vary with temperature, the variation being expressed in the usual way as the temperature coefficient.

A 5.1 V BZY88 500 mW Zener diode has a slope resistance of 75 Ω maximum at 5 mA. What would be the allowable range of currents if the diode voltage was not to vary by more than ± 0.1 V?   **Worked Example 4.4**

Solution

$$\frac{\Delta V}{\Delta I} = R_Z \qquad R_Z = 75\ \Omega \text{ and } \Delta V = \pm 0.1\ \text{V}$$

thus

$$\Delta I = \pm\ \frac{0.1\ \text{V}}{75\ \Omega} \simeq \pm 1.3\ \text{mA}$$

The allowable range of currents is thus 3.7–6.3 mA

Zener diodes are available in power ratings from 400 mW to over 20 W. Note that a Zener must be operated at a suitable current (the operating point must be beyond the 'knee' point of the *V–I* characteristic). Manufacturer's data usually states the

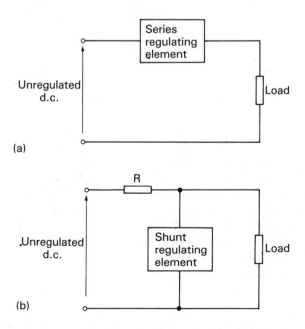

Fig. 4.9   Two arrangements of linear voltage regulators: (a) series; (b) shunt.

slope resistance at a specified current, and generally, the diode should be operated near to the stated current.

Linear Regulators

Linear voltage regulators operate by dropping an unregulated voltage through a dissipative element (either a resistor or a transistor), controlling the voltage drop so as to maintain a constant output voltage. Two possible arrangements of linear regulator are shown in Fig. 4.9. The load represents the electronic circuits to be supplied with a constant voltage. It contains active circuits, and its impedance varies with time. In the series regulator, a regulating element is placed in series with the load. The voltage drop across the regulating element is varied as the load current or the unregulated voltage varies in order to maintain a constant voltage across the load. In the shunt regulator, a resistor is placed in series with the load and a regulating element is placed in parallel with the load. The current drawn by the regulating element is varied in order to alter the voltage drop across the resistor and keep the load voltage constant. Both series and shunt regulator circuits are used in practical designs as we shall see below.

The simplest arrangement of linear voltage regulator is a shunt circuit and is shown in Fig. 4.10. The regulated voltage V_R is obtained directly across the Zener diode D_Z. When the load current I_L or the unregulated voltage V_U changes, the Zener diode current I_Z changes to compensate.

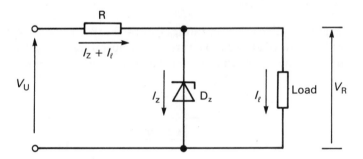

Fig. 4.10   A simple linear shunt regulator.

Worked Example 4.5   Some CMOS logic circuits operating at $+12$ V are to be included in a system and powered from a simple Zener shunt regulator circuit of the type shown in Fig. 4.10. There is a regulated $+15$ V rail available. The designer decides to use a 500 mW 12 V Zener and finds from a data sheet that the minimum recommended operating current is 5 mA. Since CMOS logic draws negligible current (micro-amperes) when quiescent, voltage regulation must be maintained down to zero load current. Select a value for the resistor and calculate the maximum allowable load current. What factors have been neglected in the calculation?

Solution

Maximum Zener current $I_{Z,\text{max}} = \dfrac{500 \text{ mW}}{12 \text{ V}} \simeq 40 \text{ mA}$

Assume load current $I_L = 0$, then $R = \dfrac{15 - 12}{0.04} = 75 \; \Omega$

$I_z + I_L$ is constant at 40 mA. I_z must not fall below 5 mA, therefore $I_{L,\text{max}} = 35$ mA.

The self-heating, slope resistance and temperature coefficients of the Zener diode have been neglected.

The simple Zener shunt circuit has several shortcomings, among which are that the Zener diode current and therefore the power dissipated in the diode vary with load current and unregulated voltage and that there is no compensation for the temperature coefficient of the diode. The simple series regulator described next partly overcomes the first of these problems. Temperature compensation of Zener diodes can be achieved by adding other devices with a similar temperature coefficient of opposite sign in series with the Zener.

Figure 4.11 shows a simple linear series regulator design using a bipolar transistor as the dissipative element. The transistor is in series with the load. The voltage at the emitter of the regulating transistor is constant at V_{BE} less than the reference voltage across the Zener diode. R_1 supplies operating current for the Zener diode and base current for the transistor. Variations on this simple circuit exist to overcome problems with temperature stability of the Zener voltage and V_{BE} drop. Protection against short circuit of the load terminals may be needed, since under these conditions the transistor passes heavy current with the full unregulated voltage across the transistor.

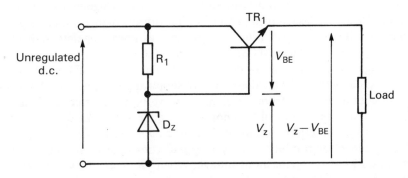

Fig. 4.11   A simple linear series regulator.

Figure 4.12 shows a simple transistor-based shunt regulator. In this design the dissipative element is the resistor R_1 (although some power is also dissipated in the shunt transistor). The shunt regulator has the important advantage of being inherently protected against a short-circuit load, because under these conditions the transistor passes no current. Another useful feature of this circuit is that it provides a path for reverse current from the load and can actually absorb power from the load. This is an advantage when the load is a d.c. motor.

Integrated-circuit voltage regulators are available for commonly used voltages such as 5 V and 15 V and with adjustable voltage outputs. They are available in both positive and negative polarities and include a voltage reference, regulating

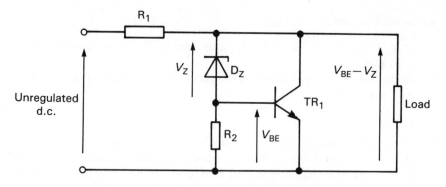

Fig. 4.12   A transistor shunt regulator.

element and control circuits within the package. They usually include current limiting and temperature compensation. Low-power types may be packaged in dual-in-line form. Higher-power devices are packaged in the same way as power transistors and require heatsinking.

Switching Regulators

Switching regulators operate at higher efficiency than linear types by avoiding power wastage in a series or shunt regulating device. They are also smaller and lighter for a given power output than linear regulators. Their disadvantage is that, because of the switching action, output ripple is usually higher than with linear regulators. Also, because of their greater complexity and higher component count, switching regulators are not usually as reliable as linear regulators. (Even so, switching regulators achieve MTBFs of around 10 000 hours.)

MTBF or *mean time between failures* is discussed in Chapter 8.

Switching regulators operate by *chopping* (switching on and off) the unregulated voltage, matching the demanded power with supplied power. A smoothing circuit produces continuous d.c. from the chopped waveform. The smoothed output is sensed and fed back to control the chopping frequency or pulse width.

Figure 4.13 shows a simple example. The circuit operates as follows. The control waveform generator produces a control signal consisting of rectangular pulses which causes TR_1 to chop the input voltage. When TR_1 is conducting current flows to the load through the inductor L_1. Capacitor C_1 is charged to the output voltage of the regulator. When TR_1 switches off, current continues to flow through L_1. The *flywheel diode D_1* is required to hold down the voltage at the collector of TR_1 and to provide a path for the current while the transistor is off. During the time that the transistor is off, the load current is being supplied from the stored energy in the inductor and capacitor. The output voltage is thus a mean d.c. level with super-imposed ripple. The mean level is regulated by comparing it with a voltage reference and generating an error signal to control the pulse generator. The output voltage may be varied by changing the frequency of the pulses, keeping their width constant, or by varying the pulse width, keeping the frequency constant. Either way, the *mark-to-space ratio* or *duty cycle* of the pulse generator output varies and controls the output voltage of the regulator.

Typical operating frequencies of switching power supplies are above 20 kHz, both to avoid generation of audible frequency acoustic noise and because inductors

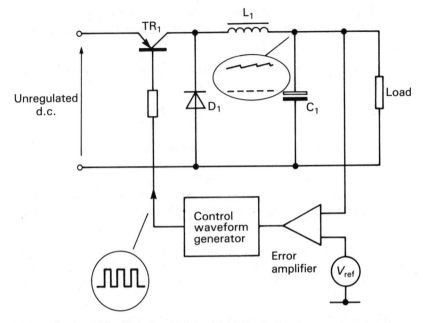

Fig. 4.13   A switching voltage regulator.

for higher-frequency operation can be made smaller and lighter than those for lower frequencies.

Summary

This chapter has looked at the main energy sources used to power electronic systems, starting with the characteristics of a.c. mains supplies and then examining energy sources such as electrochemical and photovoltaic cells which can be included within self-contained systems.

The idea of a power supply as a sub-system within an electronic system has been introduced and the performance characterization of power supplies has been discussed. The main elements of a power supply were then described, including transformers, rectifier circuits and reservoir capacitors. Voltage regulation was then introduced and the two types of regulator (linear and switching) were outlined.

The chapter has thus given a broad introduction to the subject of powering electronic equipment.

Problems

4.1   A D-size nickel–cadmium cell has a nominal 4 A h capacity and a nominal open-circuit voltage of 1.25 V. The manufacturers recommend a 12-hour charge at 500 mA. What will be the average heat dissipation during charging assuming that the cell absorbs its nominal capacity in the form of chemical energy?

4.2 The energy stored in a capacitor is $\frac{1}{2}CV^2$. Calculate the value of capacitor required to store the same amount of energy as a 30 A h 12 V car battery if the capacitor is charged to 12 V.

4.3 Why would a primary battery manufacturer advise users to store batteries in a cool place and keep them out of direct sunlight?

4.4 Recalculate (a) the capacitor value and (b) the peak rectifier current of Worked Example 4.3 for a supply frequency of 400 Hz.

4.5 What is (a) the maximum discharge rate and (b) the minimum charge rate for a secondary battery used to power a miner's helmet lamp if the lamp is to operate continuously over an 8-hour shift and then be ready for use at the start of the same shift on the following day? Assume that two thirds of the energy input to the battery is absorbed as chemical energy during charging. (*Hint*: The answers are expressed independently of the battery capacity.)

4.6 Calculate the efficiency of the circuit designed in Worked Example 4.5 at (a) maximum load current and (b) at a load current of 5 mA. Efficiency is the ratio of power delivered to the load to power drawn from the source (in this case the +15 V rail).

4.7 For the application given in Worked Example 4.5 an engineer decides to try the circuit of Fig. 4.11. He sets the Zener diode current at 5 mA by suitable choice of R_1. Calculate the efficiency for the same load currents as Problem 4.6. Neglect the transistor base current.

Passive Electronic Components 5

☐ To emphasize the differences between real components and ideal circuit elements.
☐ To introduce the properties and characteristics of real passive components.
☐ To survey the main types of passive electronic component and to discuss their fabrication.

Passive Component Characteristics

A fully detailed data sheet for an apparently straightforward component such as a capacitor contains information on a great many aspects of its behaviour. In any particular application some of the component parameters will be of the utmost importance, whilst others will be of little consequence. In some circuits, the ultimate performance which can be achieved is limited by component behaviour. Figure 5.1 shows such a circuit, a single-slope analogue-to-digital converter, or ADC. In this type of ADC a ramp generator or integrator circuit produces a ramp waveform starting at a voltage slightly below 0 V. Two comparators are used, one to compare the ramp waveform with 0 V and one to compare the analogue input voltage with the ramp waveform. As the ramp waveform increases from below 0 V to greater than the analogue input, the two comparators switch one after the other, and the time interval between the two comparators switching is accurately proportional to the difference between the analogue input voltage and 0 V. The comparator outputs are used to start and stop a counter driven from a clock circuit. The final digital value in the counter is proportional to the time interval between the two comparators switching and therefore to the analogue input voltage. Figure 5.1(b) shows the details of the ramp generator circuit, which integrates a constant reference voltage to produce a ramp. An FET switch is needed to reset the integrator at the end of each cycle by discharging the capacitor. (The control signal for the FET has been omitted from Fig. 5.1(b) for the sake of clarity.) The overall linearity of the ADC depends critically on the quality of the ramp waveform: any significant deviation from an ideal ramp, as in Fig. 5.1(c), will cause linearity errors in the ADC output. One possible cause of ramp non-linearity is the integrator capacitor: leakage current through the capacitor dielectric causes the ramp waveform to droop. A further problem is that capacitor dielectrics can absorb charge, a phenomenon which is discussed later in this chapter. The choice of capacitor for this circuit is thus very important as the linearity of the ADC depends on the capacitor.

Linearity is a very important concept in electronic engineering. A linear system obeys the principle of superposition, such that if input x_1 causes output y_1 and input x_2 causes output y_2, then an input of $x_1 + x_2$ will cause an output of $y_1 + y_2$. In the case of an ADC, linearity means that the value of the digital output is accurately proportional to the analogue input.

Explain why two comparators are used in the single-slope ADC discussed above, and why the counter is not started at the start of the ramp.

Exercise 5.1

An understanding of component parameters is essential to the circuit designer who needs to compare different types of component for a particular application. A

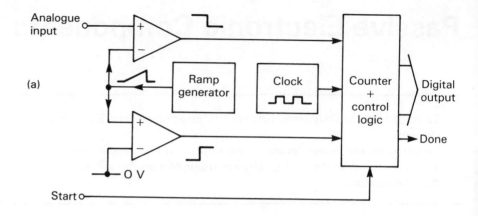

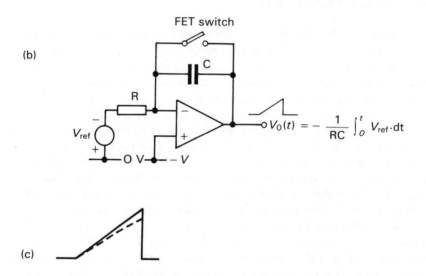

$$V_0(t) = -\frac{1}{RC}\int_o^t V_{ref}\cdot dt$$

Fig. 5.1 A single-slope analogue-to-digital converter: (a) block diagram; (b) ramp generator circuit; (c) ramp non-linearity.

good circuit designer comes to know the type of component that will be needed for a particular purpose and understands why that type of component will do the job.

This chapter starts by looking at some general aspects of passive components and then moves on to discuss specific types of component.

Tolerance

In engineering no manufactured value or dimension can ever be exact. One can, of course, approach exactness as closely as one likes, or can afford, but has to accept that a value or dimension can only ever be within some *tolerance* of the desired value. The acceptable range of a value can be specified in several ways. One is to state the upper and lower limits of the range within which the value must lie. This method is sometimes used in mechanical dimensioning, but is rarely used for component values in electronic engineering, where the normal practice is to state a *nominal* value with a *tolerance*. Often, but not necessarily, the nominal value lies in the middle of the acceptable range and the tolerance is quoted as a percentage of that nominal value. A 100 Ω $\pm 5\%$ resistor, for example, could have any value from 95 to 105 Ω. For some components, the tolerances above and below the nominal value are unequal, implying perhaps, an uneven probability distribution resulting from the fabrication process.

Electrolytic capacitors commonly have capacitance tolerances of $-10\%/+30\%$ or $-10\%/+50\%$.

The tolerances described so far have been *relative*: they represent a fractional deviation from the nominal value. For some components such as capacitors in the 1–10 pF range, the tolerance may be expressed as an *absolute* value thus: 3 pF ± 0.5 pF.

Temperature Dependence

The values of electronic components often vary with temperature because the electrical properties of materials vary with temperature. For most components, the fractional change in value is roughly proportional to the temperature change and the temperature dependence of the component value can be expressed as a temperature coefficient, often in parts per million (p.p.m.) per °C.

Worked Example 5.1

A range of $\frac{1}{4}$W resistors has a stated manufacturing tolerance of 5% and a temperature coefficient of resistance of ± 200 p.p.m. $°C^{-1}$. Ignoring any other effects on resistance value, assuming the nominal value applies at 20°C and making no assumption about the sign of the temperature coefficient, what would be the worst-case deviations from nominal value over a temperature range of 0–70°C?

Solution   The worst case values at 20°C are 1.0 $\pm 5\%$ or 0.95 and 1.05. The greatest deviation due to temperature will be at 70°C, where the change in resistance from the value at 20°C will be (70°C − 20°C) × (± 200 p.p.m. $°C^{-1}$) or $\pm 10\,000$ p.p.m. or $\pm 1\%$. Taking the worst case combinations of manufacturing tolerance and temperature dependence:

Maximum deviation = 1.05 × 1.01 = 1.06 or $+6\%$
Minimum deviation = 0.95 × 0.99 = 0.94 or -6%

Exercise 5.2

Rework Worked Example 5.1, assuming that the temperature coefficient of resistance is always negative.
(*Answers*: $+5.4\%$, -6.0%).

Stability

The electrical properties of components vary with time, whether a component is in use or in storage, due to physical and chemical changes in the materials from which the component is fabricated. Another way of looking at these changes is to say that components *age*. Component ageing can be accelerated by applied *stress*. If a component is operated continuously at its full rated voltage, for example, the component's value may change much more quickly than if the component were in storage unused. In some types of ceramic capacitors, for example, an applied voltage causes gradual changes in the crystal structure of the dielectric, resulting in a change in permittivity and hence a change in the value of the capacitor. Some other examples of stress that can accelerate ageing are: heat, which can speed up chemical changes; thermal cycling (repeatedly heating and cooling a component), which can cause joints to crack because of differential expansion; and ionizing radiation, which can disrupt the molecular and crystal structure of component materials. *High-stability* components have values that change comparatively little over time. Stability is expressed as a fractional change in value (usually in p.p.m. or %) over a stated time interval and under stated conditions.

Ionizing radiation, in the form of cosmic rays and sub-atomic particles from the Sun, is a significant problem in electronics for spacecraft.

Component ratings

Electronic components have limitations on voltage, current, power dissipation and operating temperature range. In some cases there may also be limitations that are more complicated such as rate of change of voltage. These limitations are known collectively as *ratings*. Component manufacturers usually state two sets of ratings for their products: an *absolute maximum rating*, beyond which the component will be damaged or destroyed; and a *recommended rating*, which is the manufacturer's statement of the component's capability. To say that a capacitor has a recommended rating of 16 V does not guarantee that the capacitor will work as well at 16 V as it will at 10 V: it will almost certainly be more reliable at 10 V than at 16 V, and may also be more stable. For these reasons, design engineers normally use a component well within its recommended rating.

There is, of course, a cost penalty in using a 16 V capacitor for a 10 V application, but the initial cost of the component may be less important than the long-term cost of unreliability.

Absolute maximum ratings may be important under fault or transient conditions. If a fault occurs in a system, other components will be undamaged if protective devices (such as an overvoltage trip circuit) operate before absolute maximum ratings have been exceeded.

Parasitic behaviour

So far in this discussion of passive component characteristics, we have looked at non-ideal properties which are due to the physical limitations of materials and manufacturing processes. There is another way in which passive components can be non-ideal, which is due to their electromagnetic behaviour rather than to the limitations of materials.

In lumped-parameter circuit theory there are three simple circuit elements: resistance; capacitance and inductance. Figure 5.2 summarizes their properties. Practical realizations of all three circuit elements exist in the form of resistors, capacitors and inductors, but in all three cases the practical component possesses a little of the other two circuit properties.

In a lumped-parameter circuit, we model the circuit elements as if they were localized and connected together by zero-impedance wires. Contrast this with a distributed-parameter circuit such as a transmission line where capacitance and inductance are distributed evenly along the line.

Note the suffices -ors for a component, -ance for a circuit property.

Figure 5.3(a) shows the construction of a metal film resistor, made by depositing a metallic film on the surface of a ceramic cylinder and then cutting a helical track into the film to obtain the desired resistance value. The helical construction of the resistive track suggests inductance and there is also capacitance between turns of the helix. Figure 5.3(b) suggests a possible equivalent circuit for this type of resistor, but in reality the resistance, capacitance and inductance of the component are physically distributed and not lumped as suggested. The series inductance, L_s, and the parallel capacitance, C_p, are known as *parasitic* properties of the resistor. In many applications their presence can be neglected. At low frequency, perhaps in an audio circuit, the reactance of L_s is low and the reactance of C_p is high so that the

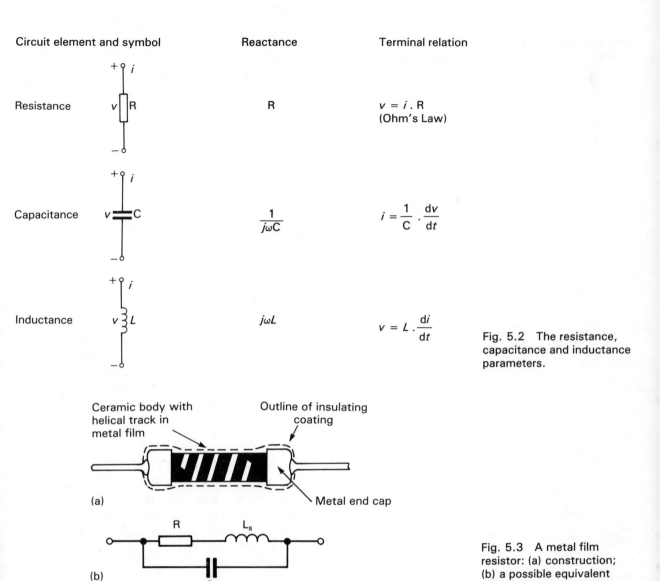

Circuit element and symbol	Reactance	Terminal relation
Resistance $\quad v \; R$	R	$v = i \cdot R$ (Ohm's Law)
Capacitance $\quad v \; C$	$\dfrac{1}{j\omega C}$	$i = \dfrac{1}{C} \cdot \dfrac{dv}{dt}$
Inductance $\quad v \; L$	$j\omega L$	$v = L \cdot \dfrac{di}{dt}$

Fig. 5.2 The resistance, capacitance and inductance parameters.

Ceramic body with helical track in metal film

Outline of insulating coating

(a)

Metal end cap

(b)

R $\quad$ L$_s$

C$_p$

Fig. 5.3 A metal film resistor: (a) construction; (b) a possible equivalent circuit.

resistor behaves almost as a pure resistance. At higher frequencies however, the impedance of the resistor is lower than at low frequencies as the reactance of C_p decreases. It is important to realize that the series inductance cannot be eliminated by making the resistor as a straight bar, although it is reduced. This is because inductance is associated with the magnetic field induced by a changing current, not with a helical or spiral conductor shape. Similarly, the parasitic capacitances are associated with the electric field between two charged conductors, not with parallel flat plates.

This point is developed further in Chapter 7.

Estimate the parasitic inductance of a metal-film resistor of the type shown in Fig. 5.3, if the ceramic body is 3 mm in diameter and the helical track consists of 10 turns spaced over a 10 mm length.

Worked Example 5.2

Compton, Chapter 8, discusses calculation of self-inductance.

Solution If we assume the ceramic has a relative permeability μ_r of 1, we can use an approximate formula for the inductance of a single-layer air-cored coil:

$$L = N^2 \mu_0 A/l,$$

where N is the number of turns in the coil, μ_0 is the permeability of free space, A is the cross-sectional area of the coil and l is its length. Hence

$$L = \frac{10^2 \times 4\pi \times 10^{-7} \times \pi \times (1.5 \times 10^{-3})^2}{10 \times 10^{-3}} \cong 90 \text{ nH}$$

Exercise 5.3 Using the parasitic inductance value calculated in Worked Example 5.2, find out at what frequency the inductive reactance of such a resistor becomes more significant than the resistance, for resistance values of (a) 10 Ω, (b) 1 kΩ and (c) 100 kΩ.
(*Answers*: (a) 18 MHz; (b) 1.8 GHz; (c) 180 GHz, which is almost into the visible region of the electromagnetic spectrum).

Figure 5.4 shows a possible equivalent circuit for a practical capacitor. The series inductance, L_s, and resistance, R_s, are due to the wire leads of the capacitor, while the parallel resistance, R_p, is due to the leakage resistance of the dielectric. If the capacitor is of wound construction, there is an additional contribution to L_s. In reality, as with the resistor discussed earlier, the parasitic properties are distributed within the capacitor to some extent. The properties of a real capacitor or inductor cannot therefore be defined analytically, and an empirical approach is taken. A capacitor or inductor is represented by a pure reactance X, and a series resistance R, both of which depend on frequency. The impedance of the component is then a function of frequency:

$$Z(\omega) = R(\omega) + jX(\omega) \tag{5.1}$$

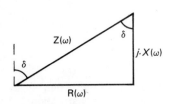

$X(\omega)$ can be recognized as the normal capacitive or inductive reactance, while $R(\omega)$ is known as the *equivalent series resistance (ESR)*. Both depend on frequency. Several other quantities are derived from R and X and all depend on frequency. The dissipation factor (DF) is the ratio of R to X and is also known as tan δ. The angle δ is called the *loss angle* and its relationship to R and X is shown by the diagram in the margin. The reciprocal of the dissipation factor is known as Q, for quality factor. Q is defined as follows:

The concept of Q factor is also used in resonant circuits and waveguide cavities with similar meanings in terms of energy losses.

$$Q = 2\pi \times \frac{\text{(Maximum energy stored in component)}}{\text{(Energy dissipated per cycle)}} \tag{5.2}$$

A capacitor or inductor with a high Q factor, or low dissipation factor, absorbs little power when used in an a.c. circuit. In tuned circuits, it is not possible to realize a high Q for the complete circuit unless the inductors and capacitors have high individual Q factors.

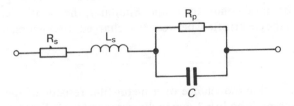

Fig. 5.4 A possible equivalent circuit for a capacitor.

Some manufacturers may state a power factor (PF), a term which is more often used in a.c. circuit theory, rather than a dissipation factor or Q factor. If the product of the r.m.s. current and r.m.s. voltage flowing into a circuit is multiplied by the PF, the result is the power dissipation in the circuit in watts. In general, the current and voltage are not in phase, so that the product of their r.m.s. values (in volt-amperes or VA) does not represent actual power dissipation.

Power factor is covered by Kip, and is an important idea in electrical power engineering.

Selection of a passive component for a particular application requires a knowledge of component characteristics in general, as introduced so far in this chapter, and also an understanding of the different types of resistors, capacitors and inductors and their fabrication and characteristics, which is what the remainder of this chapter covers.

Resistors

Resistors are used in electronic circuits for limiting current, setting bias levels, controlling gain, fixing time constants, impedance matching and loading, voltage division and sometimes heat generation. The resistance, R (Ω) of a piece of material of length l and cross-sectional area A is given by

$$R = \frac{\rho l}{A} = \frac{l}{\sigma A} \tag{5.3}$$

where ρ is the material resistivity (Ω m), σ is the material conductivity ($\Omega^{-1}\mathrm{m}^{-1}$) and l and A are expressed in metres and square metres respectively.

There are three main ways of fabricating a resistor, each applicable over a range of resistivities and resistance values. Firstly, if a material of suitable resistivity can

Table 5.1 Characteristics of resistor types

Type	Typical tolerance (%)	Typical temperature coefficient (p.p.m. °C^{-1})	Power rating (W)	Range of values	Operating temperature range (°C)	Features
Precision wire-wound	<0.5	<10	<1	0.1 Ω–20 kΩ	−55 to +150	High stability, low noise
Metal film	0.5	50	<1	0.1 Ω–1 MΩ	−55 to +125	Low noise, good stability
Carbon composition	10	1200	$\frac{1}{8}, \frac{1}{2}$, 1	10 Ω–1 MΩ	−40 to +100	General-purpose, low cost, low stability, noisy.
Carbon film	5	500	$\frac{1}{4}, \frac{1}{2}$, 1, 2	10 Ω–1 MΩ	−40 to +125	Better stability than carbon composition
Power wire-wound	5	50–250	Up to 200	0.1 Ω–10 kΩ*	−55 to +200	Higher-rated types may require heatsinks
Thick-film networks	2	300	$\frac{1}{8}$	10 Ω–100 kΩ	−55 to +125	Multiple resistors per package
Surface mount	2	200	$\frac{1}{4}$	1 Ω–10 MΩ	−55 to +125	Thick or thin film on ceramic construction, low inductance.

* Dependent on rating.

77

be made, eqn (5.3) can be realized directly in terms of a slab or rod of resistive material with metal contacts at each end. If this method of construction is not feasible, the resistor can be fabricated from a longer length of material of thinner cross-section. One way of doing this is to wind a wire onto a former, the other method is to use a film, or thin layer, of the resistive material, deposited onto an insulating substrate. Slab or rod resistors for surface mounting have the lowest series inductance of any resistor types because of the absence of leads. Wound resistors tend to have high series inductance, although this can be reduced to some extent by special winding techniques in which one part of the winding cancels the inductance of the remainder.

At high frequencies the geometric form of a resistor must be considered and special geometries such as discs may be needed.

The temperature coefficient of resistance and stability of a resistor are determined primarily by the properties of the resistive material used to fabricate the resistor. Most pure metals have temperature coefficients of resistance of around 4000 p.p.m. $°C^{-1}$, and fairly low resistivities making them unsuitable for use in resistors. Lower temperature coefficients of resistance can be achieved by fabricating resistors from special proprietary alloys with temperature coefficients as low as ± 5 p.p.m. $°C^{-1}$. Many of these alloys are based on nickel, chromium, manganese and copper. Well-known examples are the alloys known as nichrome (80% nickel, 20% chromium), constantan (55% copper, 45% nickel) and manganin (85% copper, 10% manganese, <5% nickel) with temperature coefficients of <100 p.p.m. $°C^{-1}$, <20 p.p.m. $°C^{-1}$ and <15 p.p.m. $°C^{-1}$ respectively. Metal alloys are used in the fabrication of *precision* wire-wound resistors, and metal-film resistors of the type illustrated in Fig. 5.3. These types of resistor have the lowest temperature coefficients and the best stabilities of any resistor type. High-value wire-wound resistors are not feasible because of the fairly low resistivity of resistance wire. Nichrome wire of only 0.02 mm diameter has a resistance of 3.44 kΩ m^{-1}, and as this is about the thinnest practical wire for production resistor manufacture, over 3 m of wire is required for resistor values above 10 kΩ. The range of values that can be obtained by varying the length and pitch of a helical track in a metal film is limited and higher-resistance values require either thinner films or films of higher sheet resistivity, which can be obtained by including non-conducting materials in the film during deposition from a vapour.

A sheet of conductor has a resistivity, measured in ohm per square (Ω $\square^{-1}$), which is a constant for any sized square of the sheet. Metal films can be fabricated with sheet resistivities of up to about 20 kΩ $\square^{-1}$.

For *general purpose* use where low temperature coefficients and high stability are not essential, cheaper resistors can be fabricated using carbon as the resistive material. Carbon composition resistors are the cheapest type of resistor for general-purpose electronic use and are fabricated from carbon granules mixed with a binder and moulded into a rod with lead wires at each end. Carbon film resistors are more stable than carbon composition types and are only slightly more expensive. They are fabricated by pyrolytic decomposition of a carbon-containing gas, such as methane, in a furnace, depositing a carbon film onto a ceramic or glass substrate. The resistors are then fitted with end caps and leads and coated with a protective varnish, lacquer or plastic.

A third type of resistive material used in resistor fabrication is known generically as *cermet*, for ceramic–metal. These materials contain finely divided metals distributed in a glassy vitrified ceramic which can be printed or painted onto a substrate in the form of a paste and then *fired* to fuse the constituents into a hard solid. There are two important applications for these materials. One is in variable resistors or potentiometers where a low temperature coefficient of resistance

(<200 p.p.m. °C^{-1}) is required. The other is in *thick-film hybrid circuits* and thick-film resistor networks. Thick-film circuits consist of resistors and small-value capacitors printed and fired onto a ceramic substrate. Surface-mounted ICs and transistors can be soldered to the circuit, which can then be coated with epoxy-resin and used either as a complete self-contained circuit or as a component on a conventional PCB. Thick-film resistor networks are made in the same way but contain only resistors. They are especially useful where several resistors of the same value are required in the same location on a PCB. A common application is a group of eight *pull-up* resistors connected to a microprocessor bus with a common connection to a supply rail. A thick-film single-in-line (SIL) resistor pack containing eight commoned resistors has only nine terminals and occupies a smaller board area than eight discrete resistors.

Thick-film circuits are discussed in more detail by Till and Luxon.

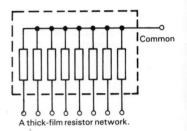

A thick-film resistor network.

Noise

All resistors generate electrical *noise* or small random fluctuations of voltage or current. Noise is not necessarily due to material imperfections in resistors, although some types of resistor are more noisy than others: noise is a fundamental property of all conductors. Electronic systems handling weak signals have to be carefully designed to minimize the amount of noise added to the signal during processing and to achieve an acceptable *signal-to-noise ratio*. There are two types of noise generated in resistors: *Johnson noise*, which occurs in all resistors, and *excess noise* which occurs in some types of resistor. An additional type of noise, *shot noise*, is important in devices with a potential barrier such as p–n junction diodes and junction transistors but does not occur in resistors. Johnson noise is due to the random thermal motions of electrons in a resistor. The r.m.s. noise voltage developed in a resistor of value R at an absolute temperature T is given by:

Senturia and Wedlock discuss noise in more detail.

$$\sqrt{(\overline{v_n^2})} = \sqrt{(4\ kTRB)} \tag{5.4}$$

where k is Boltzmann's constant (1.38×10^{-23} J K^{-1}) and B is the bandwidth or frequency range over which the noise is measured. Johnson noise has uniform spectral density at all frequencies and is described as *white noise* by analogy with white light.

There is an upper frequency limit set by quantum effects. Eqn (5.4) is valid for frequencies below f where $kT \ll hf$ and h is Planck's constant (6.626×10^{-34} J s). At room temperature (300 K) eqn (5.4) is valid to more than 500 GHz.

What is the r.m.s. noise voltage developed in a 1 MΩ resistor at room temperature (290 K) over a 20 kHz bandwidth due to Johnson noise?

Worked Example 5.3

Solution   From eqn (5.4),

$$\sqrt{(\overline{v_n^2})} = \sqrt{(4 \times 1.38 \times 10^{-23} \times 290 \times 10^6 \times 20 \times 10^3)}$$
$$= 18\ \mu\text{V r.m.s.}$$

Johnson noise depends only on resistor value and temperature and is independent of the resistor material. Wire-wound resistors exhibit only Johnson noise. Other resistor types generate *excess noise* over and above the inevitable Johnson noise. Excess noise increases with current. Carbon composition resistors are the noisiest type.

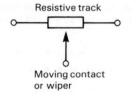

Resistive track

Moving contact
or wiper

Potentiometers

Resistors with a movable contact or tap are called potentiometers and find application in electronic circuits as operating adjustments or as manufacturing adjustments (*trimmers* or *presets*). Special potentiometers are also used as angular position transducers in servo-mechanisms. The resistive medium can be a wire winding or a cermet or carbon film, with the same characteristics, such as temperature coefficient, as fixed resistors of the same material. The potentiometer track can be straight with a sliding contact, or circular or helical with a rotating contact. Helical tracks are used on multi-turn potentiometers for applications where fine adjustment of resistance is required. The potentiometer track may be of uniform resistance (linear) or with resistance varying with position, usually logarithmically. Power ratings are generally less than 1 W.

Capacitors

Capacitors are used in electronic circuits for storing charge, as elements of frequency-selective circuits and filters, for coupling a.c. signals from one circuit to another and for shunting unwanted signals to ground (decoupling). Most capacitors used in electronic engineering are of the parallel-plate type whose capacitance, C in farads (F) is given by

A farad (F) is 1 coulomb per volt. The capacitance of a capacitor, C, charged to V volts by a charge Q coulombs is given by $C = Q/V$.

$$C = \epsilon_0 \epsilon_r A/d \tag{5.5}$$

where A is the area of the plates (m²), d is the plate separation (m), ϵ_0 is the permittivity of free space and ϵ_r is the relative permittivity of the dielectric medium between the plates (dimensionless). An alternative way of quantifying the multiplying effect of the dielectric is to state the dielectric constant, K, of the material where $K = \epsilon_0 \epsilon_r$. The properties of dielectrics have a significant influence on the properties of capacitors and will be briefly described.

Dielectrics

Since $\epsilon_r \simeq 1.0006$ for air, we can regard air as almost equivalent to a vacuum in terms of permittivity.

Polarization is discussed by Kip, Kraus and Carver, Carter, and Compton. A more detailed discussion of polarization and dielectrics is given by Anderson *et al.*

Eqn (5.5) shows that the presence of a dielectric between the plates of a capacitor increases the capacitance by a factor of ϵ_r over the same capacitor with air between the plates. The increased capacitance due to the dielectric is caused by *polarization* of the dielectric in the presence of an electric field. Polarization is a slight shift of the negative charge in the dielectric relative to the positive charge which causes a net charge to appear on opposite faces of the dielectric. In a capacitor the polarization charges neutralize some of the free charge on the capacitor plates and reduce the potential difference between the plates. Since capacitance is defined as Q/V, a reduction in the potential difference, V, implies an increase in capacitance. Put another way, because the polarization charges neutralize some of the free charge stored on the plates, the capacitor is able to accept more stored charge for a given potential difference between the plates.

There are several different mechanisms at the atomic or molecular level which can contribute to the polarization and therefore the relative permittivity of a dielectric. Each mechanism has its own characteristic time constant and the relative permittivity of a dielectric is therefore frequency dependent. The slowest polariza-

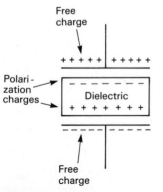

Free
charge

$+ + + + +$ $+ + + + +$

Polari-
zation
charges

Dielectric

$+ + + + + + +$

Free
charge

tion mechanism which contributes to the relative permittivity of a dielectric at low frequencies only is the migration of free electrons and impurity ions in the presence of an electric field. There are two polarization mechanisms at molecular level which can operate in materials formed from or containing polar molecules. One is orientation of the polar molecules in the direction of the applied field, the other is stretching or distortion of the polar molecules in the direction of the field. The fastest polarization mechanism, which operates at optical frequencies, is displacement of the electron clouds in atoms of the material relative to their nuclei.

In a polar molecule there is an asymmetric distribution of positive and negative charge, and the molecule has a dipole moment.

When a dielectric is polarized by an alternating electric field energy losses occur in the dielectric. These energy losses generate heat and are responsible for the resistive component $R(\omega)$ of the complex impedance of a capacitor in eqn (5.1). The loss angle, δ, of a capacitor can be seen to be a property of the dielectric.

Polymer films used in capacitor manufacture have relative permittivities between 2 and 5. Metal oxides can have values as high as 80 (titanium dioxide) and some ceramics, notably barium titanate can have relative permittivities of up to 12 000. The latter are known as *ferroelectric* or *high-K* ceramics and although they can be used to fabricate compact capacitors, the capacitance values are not very stable.

Ferroelectric materials are discussed by Anderson *et al*.

Table 5.2   Characteristics of capacitor types

Type	Typical tolerance[1] (%)	Typical temperature coefficient (p.p.m./°C)	Maximum voltage[2] (V)	Range of values[2] (F)	Operating temperature range (°C)	Features
Silvered mica	1	50	2500	1p–10n	−40 to +85	High Q, high stability, high cost.
Ceramic	10	100–750	500	1p–1μ	−55 to +125	
Polystyrene	2	<200	500	10p–10n	−40 to +70	Low leakage (>10^{12} Ω)
Polycarbonate	5	50	500	10n–10μ	−55 to +100	
Polyester	10–20	200	500	1n–1μ	−55 to +100	
Polypropylene	20	200	1500	1n–1μ	−55 to +100	Low leakage (>10^{11} Ω)
Solid tantalum	10–20	—	35	10n–300μ	−55 to +85 or +100	Polarized. Leakage depends on value.
Aluminium electrolytic	20	—	400	1μ–1000μ	−40 to +85	Short life, wide temperature variation.
Al. elect. computer grade	+80/−20	—	400	Up to 150 000μ	−40 to +85	Longer life, high capacitance.

[1] An absolute tolerance often applies for values below 10 pF.
[2] High values and high maximum voltage are not available simultaneously.

The maximum voltage that a capacitor can withstand without damage is determined by the dielectric strength of the dielectric (in V m^{-1}) and by the thickness of the dielectric. For a given dielectric material, capacitors of higher working voltage can be made by using a thicker dielectric, but as eqn (5.5) shows, increasing the dielectric thickness, d, decreases the capacitance. The physical size of capacitors therefore increases with capacitance value and with working voltage.

Dielectrics are not perfect insulators and some current will flow between the plates of a capacitor with a steady voltage applied. Capacitors therefore, have a d.c. *leakage resistance*, which for the best capacitor types is as high as 10^{12} Ω for a capacitance of around 10 nF. Apart from a steady leakage current, capacitors also exhibit *dielectric absorption* in which charge is absorbed into the dielectric. A capacitor which has been short-circuited and supposedly fully discharged can recover a small voltage as the absorbed charge emerges from the dielectric. This can be a problem in precision analogue-to-digital converters, where the result of an A-to-D conversion can be slightly influenced by the result of the previous conversion because the capacitor has some 'memory' of its previous voltage.

Capacitors can be divided broadly into two types: those that have a solid dielectric of ceramic or polymer, and those that have a thin metal-oxide film as a dielectric. For reasons that will be seen later, the latter are known as *electrolytic capacitors* and for want of any better name we can refer to the remaining types as *non-electrolytic*.

Leakage resistance less than this can be created along the outside of a capacitor by contamination with flux or even oils from the human skin.

Non-electrolytic Capacitors

Non-electrolytic capacitors used in electronic engineering fall into three main groups: mica, ceramic and polymer film. Mica capacitors are fabricated from thin sheets of naturally occurring mica, a laminar aluminium silicate which can be split into sheets as thin as 25 μm. The capacitor plates are formed from a silver coating on each side of the sheet. Silvered mica capacitors are stable and have a high Q factor. They are used for this reason in resonant circuits. Because mica cannot be rolled up there is a practical limit of about 100 nF on capacitor value and except in small values, multiple sheets of mica have to be stacked together to obtain the required capacitance.

Exercise 5.4 Estimate the maximum capacitance value achievable by stacking sheets of mica 25 μm thick measuring 30 mm square if a maximum of ten sheets can be stacked. ϵ_r for mica is about 6 and ϵ_0 is about 9 pF m^{-1}.
(*Answer:* 20 nF).

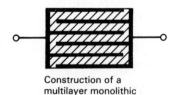

Construction of a
multilayer monolithic
ceramic capacitor

Ceramic capacitors are made either as flat discs or as rectangular multi-layer monolithic blocks depending on the value required. The ceramics used are mainly based on titanium dioxide (TiO_2) or barium titanate ($BaTiO_3$). Other compounds are added to the ceramic to give specific electrical properties. Dielectrics based on barium titanate have very high relative permittivities and are known as high-K dielectrics. High-K ceramic capacitors have poor stability and are thus best suited to applications such as decoupling, where a stable capacitance value is not essential but where the compactness of a high-K capacitor is useful. Ceramic capacitor electrodes are formed from silver, platinum or palladium which is normally applied as a paste and then fired to fuse with the ceramic. If leads are required they are then soldered to the electrodes, and the capacitor body is coated to protect the solder joints. Ceramic 'chip' capacitors without leads for surface mounting need no protective coating and give excellent high-frequency performance because of the lack of leads and the consequent low series inductance.

Polymer film capacitors are made either by coating a plastic film with metal by vapour deposition or by forming a sandwich of metal foil and plastic film. The film or foil is then rolled into a cylinder with either axial or radial leads. Polymer capacitors tend to be bulky for their capacitance because of the low relative permittivity of polymers. A variety of polymers are used in capacitor construction, of which only the most common are included in Table 5.2.

Electrolytic Capacitors

As we have seen from eqn (5.5), large-value capacitors must have a dielectric of high relative permittivity and minimum thickness together with a large plate area. Electrolytic capacitors combine all of these factors and achieve the lowest volume for their capacitance and voltage rating, of any capacitor type. They are used in circuit applications requiring large values of capacitance, such as power-supply circuits where they find use as reservoir capacitors, low-frequency filters, multistage amplifiers, and timing circuits with long time constants.

The construction of a typical electrolytic capacitor is shown in Fig. 5.5. The capacitor plates are metal foil and the dielectric is a thin film of metal oxide formed on one of the plates, called the anode. The dielectric is connected electrically to the other plate (the cathode) by an electrolyte or ionic conductor. The cathode and

The charge carriers in an electrolyte are mobile *ions* or ionized atoms which have either gained or lost one or more electrons. An electrolyte need not be liquid.

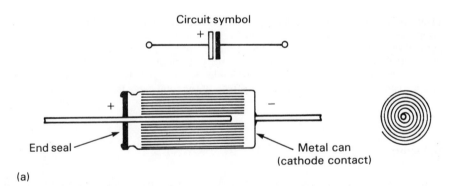

Circuit symbol

(a)

End seal

Metal can (cathode contact)

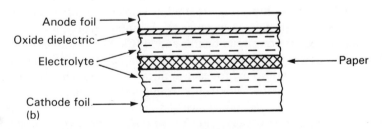

Anode foil
Oxide dielectric
Electrolyte
Paper
Cathode foil
(b)

Fig. 5.5 Construction of an electrolytic capacitor: (a) general arrangement; (b) detail of foils, dielectric and electrolyte.

dielectric are also separated by porous paper sheets to prevent direct contact between the cathode and dielectric. The paper does not prevent current flow because of its porosity. The type of electrolytic capacitor just described is *polarized* and must be connected in a circuit with the anode at a more positive potential than the cathode. This type of electrolytic capacitor is the most common and has the greatest capacitance for its volume of all electrolytic types. If a polarized electrolytic capacitor is subjected to reverse polarity either the dielectric will break down or an oxide film will form on the cathode, reducing the capacitance. Non-polarized electrolytic capacitors are manufactured with oxide films on both electrodes for a.c. circuit applications. Their capacitance is lower than for a polarized type of the same volume, for the reason just mentioned.

An oxide film on both electrodes is equivalent to two capacitors in series. Electrolytic capacitors may explode if subjected to reverse polarity.

The two most common electrolytic capacitor types are made from either aluminium or tantalum. Aluminium electrolytics consist of two aluminium foil electrodes, one of which has an aluminium oxide film formed on its surface, separated by a wet or paste electrolyte. They are characterized by a wide tolerance on initial value, considerable capacitance variation with temperature and a short life (less than a year) at high temperatures. Their capacitance decreases with age. Special long-life grades of aluminium electrolytic capacitors are made for use in computer power supplies. Tantalum electrolytic capacitors can be made with foil electrodes and wet electrolyte in the same way as aluminium electrolytics. They can also be made with a solid dry electrolyte, achieving long life and high reliability at reasonable cost. The dielectric film is tantalum pentoxide, formed in a thin layer and with large surface area by sintering tantalum powder into a solid slug or bead anode which is coated with carbon and then metal-plated to form the cathode contact.

Exercise 5.5

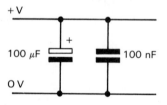

+V

100 μF   +   100 nF

0 V

Electrolytic capacitors are often used for power rail decoupling on PCBs to shunt any interference present on the rail. Often, circuit designers add a small ceramic capacitor as shown in parallel with the electrolytic. The value of the ceramic capacitor might be only 100 nF while the electrolytic could be of 100 μF or more. Capacitances in parallel add, so the total capacitance of the combination shown would be 100.1 μF. What is the purpose of the ceramic capacitor despite its negligible effect on the total capacitance?

Inductors

Inductors are still referred to as *chokes* in some applications.

Although transformers are not discussed here, much of what follows about inductor manufacture is also relevant to the fabrication of small transformers for signal coupling, isolation and impedance matching.

In most low-frequency electronic circuits, inductors are usually avoided altogether because of the expense and inconvenience of manufacture and because of the better control over circuit characteristics obtainable by using capacitors as the reactive elements. At higher frequencies inductors are used as elements of tuned circuits and filters. At microwave frequencies inductance can be obtained at minimal cost by inductive conductor patterns on circuit boards and in ICs. Power inductors are used in switched-mode power supplies as described in Chapter 4. Most inductors have to be specially designed and wound for each application, although there are a few off-the-shelf inductors available with values from 1 μH to 1 mH.

There is no simple equation for the inductance of an inductor analogous to eqn (5.5) for parallel plate capacitors because inductor geometries vary considerably depending on the method of manufacture. For the *special case* of an *N*-turn inductor wound on a toroid of a suitable material the inductance is

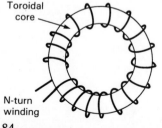

Toroidal core

N-turn winding

$$L = N^2 \mu_0 \mu_r \frac{A}{l} \qquad (5.6)$$

where μ_0 is the permeability of free space ($4\pi \times 10^{-7}$ H m^{-1}), μ_r is the relative permeability of the toroid material (dimensionless), A is the cross-sectional area of the toroid (m^2) and l is the mean circumference of the toroid (m). The toroid forms a magnetic circuit which contains the magnetic flux generated by a current in the winding. In other geometries the magnetic flux is not totally contained within the magnetic circuit and eqn (5.6) does not apply. Nevertheless it does give some insight into inductors in general. Inductance is proportional to the square of the number of turns, N, in the winding, whereas the resistance, volume and weight of the winding are proportional to N. The cross-sectional area, A, of the magnetic circuit should be large to obtain a large value of L, but the length of the magnetic circuit, l, should be kept small. Small, fat toroids are preferred therefore (but with only a small hole through the middle there will not be much space for windings). As in the case of capacitors where a dielectric of high relative permittivity is desirable to obtain large values of capacitance, in inductors the core material should have high relative permeability. Ferromagnetic metals such as iron, nickel, cobalt and their alloys have values of μ_r as high as 200 000. Inductors with metal cores tend to have high losses because of circulating *eddy currents* in the core. Low-loss inductors can be fabricated using *ferrites* as the core material, which have high μ_r but are electrically insulating. Ferrite components for inductors are widely available commercially. Manufacturers give detailed empirical data on inductor design using their components. The smallest ferrite components are *ferrite beads* which are designed to thread onto a wire to form a single-turn inductor. They are used to control parasitic signal pickup and to filter unwanted high-frequency interference.

Compton and Carter discuss calculation of inductance in more general cases.

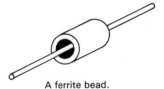

A ferrite bead.

Relays

Electromagnetic relays are still used in some application areas of electronics. They consist of a coil with an iron core and a movable iron pole-piece which operates one or more sets of switch contacts. The coil is isolated from the switch contacts so that a low-power circuit driving the coil can switch an independent circuit, possibly carrying much larger currents or voltages. In many electronic applications *reed relays* are often used. They consist of a *reed switch* in a glass envelope, fitted axially into a solenoid as shown in Fig. 5.6(b). Reed relays are capable of fast switching, and can often be driven directly by logic circuits. A relay coil is, of course, an inductor and the transistor circuit shown in Fig. 5.6(c) includes a *freewheel diode* to provide a path for the current generated by the coil as the magnetic field collapses after the transistor has turned off.

Summary

In this chapter the differences between real components and ideal circuit elements have been examined, concentrating on passive components. All passive components are subject to manufacturing variations which are quantified by an engineering tolerance on a component's value. Component values vary with temperature and with age. Temperature variation is normally expressed using a

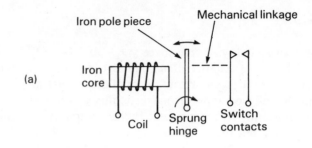

(a)

Iron pole piece

Mechanical linkage

Iron core

Coil

Sprung hinge

Switch contacts

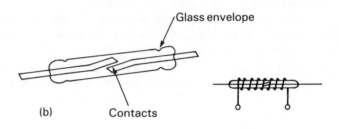

Glass envelope

(b)

Contacts

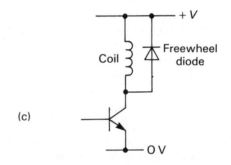

(c)

+ V

Coil

Freewheel diode

0 V

Fig. 5.6 Electromagnetic relays: (a) conventional type; (b) reed type; (c) transistor drive circuit.

temperature coefficient and the effect of age is expressed as a component's stability.

All components have parasitic properties due to their electromagnetic behaviour and their materials. The parasitic properties include unwanted reactance and resistance, energy losses in capacitors and inductors and noise in resistors.

The main types of resistor and capacitor used in electronics have been introduced and inductors have been discussed briefly.

Problems

5.1 The voltage gain of a particular precision amplifier is set by the ratio of two resistors and must be accurate to within $\pm 2.5\%$ of the design value over the full operating temperature range of the amplifier (0–40°C) and over the full expected life of 5 years. The designer chooses metal film resistors with a temperature coefficient of ± 50 p.p.m. $°C^{-1}$ and with a stability of $\pm 1\%$ of initial value over 5 years. What initial tolerance on the resistor values at 20°C is needed?

5.2 If a 1 kΩ resistor is to be used at 50 MHz, what is the maximum acceptable parasitic shunt capacitance of the resistor if the capacitive reactance is to be no less than ten times the resistor value?

5.3 Low-leakage capacitors can have insulation resistances of 10^{12} Ω. What is the r.m.s. noise voltage developed in a resistance of this value at 300 K over a 10 kHz bandwidth? Why are practical capacitors not noisy?

6 Heat Management

Objectives
- ☐ To introduce the physical principles of heat and common solutions to heat problems in electronic systems.
- ☐ To review heat-transfer mechanisms in the context of electronic systems.
- ☐ To introduce heatsink technology and simple heatsink calculations.
- ☐ To discuss briefly some heat-removal methods used in high-density electronic systems.

Heat is generated in all operating electronic systems as a consequence of the flow of electric current. Heat is energy (in the form of molecular or atomic vibrations) and is measured in the same units (joules) as any other form of energy. Rate of generation or transfer of heat is therefore expressed in joules per second or watts.

Heat generation in a conductor carrying electric current arises from interactions between the charge carriers and the atoms within the conductor, resulting in the transfer of some energy from the charge carriers to the atoms of the conductor. At circuit level, the concept of electrical resistance is used to explain the heating effect of an electric current. Heat can also be generated in parts of an electronic system where current flow is either undesired or unexpected. Examples include: eddy currents circulating in the magnetic core of a transformer or inductor causing heating of the core; heating in the walls of a waveguide due to induced currents caused by the propagation of electromagnetic waves through the guide and heat generated in the dielectric of a capacitor whilst charging and discharging.

The most obvious and extreme effect of heat generation in electronic components is thermal destruction: most electronics engineers have accidentally destroyed components during testing of breadboard or prototype circuits by applying a power-supply voltage which is too high or by inadvertently causing a short circuit, resulting in excessive heating and consequent destruction.

Rated operating temperatures of electronic components can be surprisingly high. Military-grade integrated circuits and transistors, for example, can be operated at ambient temperatures of 125°C, while some power resistors can operate at over 150°C. Despite this apparent robustness, most components in commercial and military designs are operated at much lower temperatures by conservative choice of device rating or by taking special measures to remove heat. The reason for this is that *reliability* decreases rapidly with increasing temperature because higher temperatures accelerate many physical and chemical processes such as diffusion, material creep and corrosion which can contribute to component deterioration and eventual failure. Repeated temperature changes (thermal cycling) have an even more drastic effect on component life, because differential expansion within a component induces mechanical stress which can lead to cracking of materials or separation of bonded joints.

Heat generation in a resistance R due to an imposed voltage V or a current I is given by V^2/R or I^2R or VI. These equations are valid for steady voltages and currents. In a.c. circuits, phasor multiplication and division must be used.

Dielectric losses in capacitors were discussed in Chapter 5.

If a resistor is required to dissipate $\frac{1}{8}$ W in a circuit application, a conservative choice of component might be a $\frac{1}{4}$ W or $\frac{1}{2}$ W resistor, either of which would operate at a lower temperature than a $\frac{1}{8}$ W device, while dissipating $\frac{1}{8}$ W.

Reliability is defined in Chapter 8 and the effect of temperature on reliability is discussed in Chapter 9.

Heat Transfer

There are three main physical mechanisms by which heat can be transferred: conduction, convection and radiation. The first two of these are the most important in electronic engineering, although radiation can contribute to heat dissipation from a blackened heatsink. One important exception to this general statement, however, is in satellite design, where radiation is the only mechanism by which heat can be lost ultimately from the craft.

For more information on heat transfer see Wong.

Thermal Conduction

Conduction is important in solid materials such as silicon, aluminium, copper and plastic. Good electrical conductors are also good thermal conductors. Conversely, electrical insulators are poor conductors of heat. Thermal conduction is the first process involved in removing heat from the interior of an electronic component.

Both electrical and thermal conductivity depend upon electron mobility and good electrical conductors (metals) are also good thermal conductors for this reason.

The ability of a material to conduct heat is quantified by the thermal conductivity of the material, k, expressed in W m^{-1} °C^{-1} and defined by the expression:

$$\frac{dQ}{dt} = -kA \frac{d\theta}{dx} \tag{6.1}$$

Thermal conductivity is properly quoted in W m^{-1} K^{-1}, but since temperature difference rather than absolute temperature is used, °C has been used here. This is also consistent with the use of °C in stating thermal resistance values.

where dQ/dt is the rate of heat flow (W) through a cross-sectional area A of the material (m²), and $d\theta/dx$ is the temperature gradient (°C m^{-1}) normal to the area.

Convection

Convection is an important heat transfer mechanism in fluids (gases and liquids), and is the means by which heat is ultimately removed from most electronic systems, usually by transfer of heat to the surrounding air. Convective heat transfer from a hot surface to the nearby air is shown in the margin. The air immediately adjacent to the hot surface becomes warmed by conduction and decreases in density as a result of the increase in temperature. The warmed air is then displaced by cooler air of higher density. The displaced warm air, of course, carries heat away with it. Convective heat transfer can be enhanced by increasing the area of the hot surface (by adding fins), by placing the hot surface vertically, and by forcing cooling air across the surface (forced convection). Convective heat transfer is governed by the same form of expression as was given above for conduction:

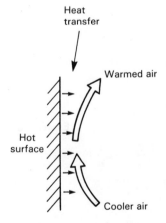

$$\frac{dQ}{dt} = hA \, \Delta\theta \tag{6.2}$$

This equation is known as Newton's law of cooling.

where dQ/dt is the rate of heat loss (W) from an area A (m²). $\Delta\theta$ is the temperature difference between the hot surface and the surrounding fluid (°C). h is then a convective heat transfer coefficient (W m^{-2} °C^{-1}). Unfortunately, the coefficient h depends on temperature and is not easily calculated. As we shall see below, however, for simple heat-removal design problems, we can use the concept of *thermal resistance*.

Radiation

Hot surfaces, even in a vacuum, can lose heat by emission of electromagnetic radiation. At the temperatures encountered in electronic systems (say $-50°C$ to $150°C$), radiated energy will be in the microwave and infra-red regions of the electromagnetic spectrum. Heat loss by radiation is governed by the expression:

$$\frac{dQ}{dt} = \sigma \epsilon A(\theta_s^4 - \theta_A^4) \tag{6.3}$$

where dQ/dt is the rate of heat loss (W) from an area A (m²), σ is the Stefan–Boltzmann constant (5.67×10^{-8} W m⁻² K⁻⁴), θ_s is the temperature of the emitting surface (K) and θ_A is the temperature of the surroundings (K). ϵ is the emissivity of the hot surface (dimensionless and ≤ 1).

This expression implies assumptions about the natures of the emitting surface and the surroundings. See Wong for more information.

Thermal Resistance

Practical heat calculations are simplified by using the concept of *thermal resistance*, which has the units of °C W⁻¹. For a general heat transfer problem:

$$\Delta\theta = \frac{dQ}{dt} R_\theta \tag{6.4}$$

where $\Delta\theta$ is a temperature difference (°C), dQ/dt is heat transfer rate (W) and R_θ is a thermal resistance (°C W⁻¹). This equation is analogous to Ohm's law with temperature replacing voltage, heat transfer rate replacing current and thermal resistance replacing electrical resistance. The thermal resistance can take account of conduction, convection and radiation effects provided we use the thermal resistance value only as an approximation and only over a limited range of temperatures.

In most applications, the thermal resistance values needed can be found in manufacturer's data sheets and there is no need to perform calculations from first principles.

Exercise 6.1 Calculate the thermal resistance between the end faces of a 10 mm × 10 mm aluminium alloy bar of length 100 mm. The thermal conductivity, k, of the alloy is 170 W m⁻¹ °C⁻¹. Ignore heat loss from the sides of the bar.
(*Answer*: about 6°C W⁻¹).

Heatsinking

The term *heatsink*, meaning a device for conducting heat from a power semiconductor and dissipating that heat to the surroundings, is poor terminology: the true heat *sink* is the surrounding environment.

Heatsinks are used to conduct heat from power semiconductors and resistors, and to dissipate this heat to the surroundings. The main heat-dissipation mechanism is convective heat transfer to the surrounding air, although radiation can also contribute to the heat loss. Many heatsinks have a blackened surface finish (which adds negligible cost to the heatsink) to enhance radiated energy loss. For higher-power dissipations, fan-assisted removal of heat from heat sinks is common. Small power devices dissipating up to about 2 W may be cooled by clip-on heat dissipators. Figure 6.1 shows a selection of heatsinks and heat dissipators. All of

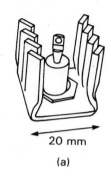

20 mm

(a)

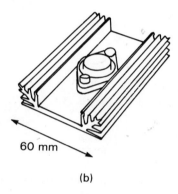

60 mm

(b)

16 mm

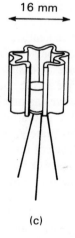

(c)

Fig. 6.1   Heatsink and transistor cooler designs: (a) 20°C W^{-1}; (b) 2–5°C W^{-1} depending on length; (c) 30–50°C W^{-1} depending on length. (Designs illustrated courtesy Redpoint Ltd)

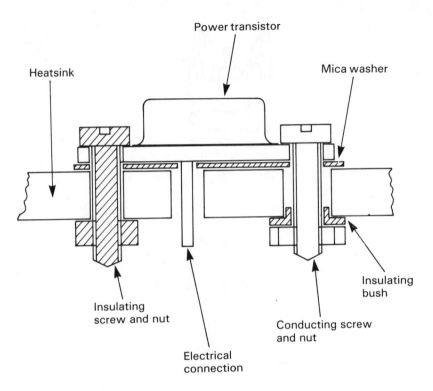

Fig. 6.2   Mounting arrangements for a power transistor.

Power transistor

Heatsink

Mica washer

Insulating
screw and nut

Electrical
connection

Conducting screw
and nut

Insulating
bush

these devices are made from aluminium alloy because of its high thermal conductivity and the ease with which it can be extruded to form the complex finned shapes needed for effective convective heat transfer.

Designers of power semiconductors pay careful attention to the thermal design of their packages to ensure the lowest possible thermal resistance between the power dissipating regions of the device and the outer mounting surface of the case. In many instances the metal case serves as one of the electrical terminals. Quite often the case terminal is not at 0 V potential and the power device has to be insulated from the heatsink. Unfortunately, electrical insulators are also good thermal insulators and only a thin layer of insulator can be allowed if the thermal resistance between the case and the heatsink is to be kept low. Thin washers of mica (a naturally occurring laminated mineral) are often used and are available to suit all types of power semiconductor package. Insulating screws (of nylon) or insulated bushes for use with metal screws are needed to fasten a power device to a heatsink. Figure 6.2 shows a typical power transistor mounted on a heatsink and illustrates the alternative fastening methods.

To improve heat transfer from the case of a power semiconductor through a mica washer to a heatsink, a heat-conducting compound such as silicone grease is smeared onto both sides of the washer before assembly. The grease fills voids left between the metal-to-washer interfaces which would otherwise be filled with air, which is a poor thermal conductor and would increase the overall thermal resistance.

Some device data sheets do not state thermal resistance directly: instead they

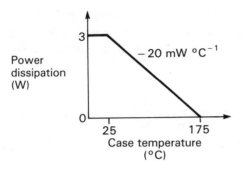

Fig. 6.3   A typical power-dissipation derating curve for a small silicon transistor.

may include a power-dissipation derating curve similar to that shown in Fig. 6.3 or an equivalent statement in words. The thermal resistance of the device is the reciprocal of the slope of the derating curve (the thermal resistance is, of course, always positive).

Steady-state Heatsink Calculations

The simplest heatsink situation to handle is steady-state power dissipation with one power device on a heatsink. (The same calculation applies for a small power device with a clip-on heat dissipator except that there would be no mica washer.) Figure 6.4 shows the electrical analogue of the thermal path. The variables in this arrangement are: the power dissipation, P (W); the thermal resistances, R, (°C W^{-1}) which are all in series; and the two temperatures, θ_A (the ambient temperature of the surroundings) and θ_J (the internal junction temperature of the power device). These quantities are related by the thermal analogue of Ohm's law:

$$\theta_J - \theta_A = P(R_{J\text{-}C} + R_{C\text{-}S} + R_{S\text{-}A}) \qquad (6.5)$$

Of these quantities $R_{J\text{-}C}$ and $R_{C\text{-}S}$ are often known from the choice of power device, while various combinations of the remaining values are to be decided. As with many design problems, the final choice of heatsink or allowable power dissipation may be arrived at after several iterations.

Worked Example 6.1

A voltage regulator IC dissipating 25 W is to be mounted on a heatsink such that the junction (internal) temperature of the IC is to be limited to 125°C. What is the maximum allowable thermal resistance of the heatsink if the thermal resistance of the regulator (junction–case) is 1.7°C W^{-1}, the thermal resistance of the mica washer and silicone grease is 0.3°C W^{-1}, and the ambient temperature is 25°C?

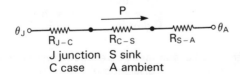

Fig. 6.4   Electrical analogue circuit for a power device on a heatsink (steady state).

Solution   Total thermal resistance allowable = $(125 - 25°C)/25 \text{ W} = 4°C \text{ W}^{-1}$
Thermal resistance, junction–case and mica washer is $1.7 + 0.3 = 2°C \text{ W}^{-1}$.
Therefore the maximum allowable thermal resistance of the heatsink is $4 - 2 = 2°C \text{ W}^{-1}$.

Exercise 6.2   For the conditions of Worked Example 6.1, if the heatsink actually used had a thermal resistance of $1.5°C \text{ W}^{-1}$, what would the junction temperature of the IC be?
(*Answer*: 113°C)

Transient or Dynamic Heatsink Calculations

We have seen in the previous section, how to select a heatsink to dissipate heat from a power device with a steady power dissipation. Worked Example 6.1 has suggested an application where such a calculation might be used. In other applications, however, power dissipation may not be steady: it may fluctuate unpredictably (as in an audio power amplifier) or may be pulsed with a regular and predictable frequency and waveform. Sometimes, it may be necessary to design for a worst-case maximum continuous power dissipation using the steady-state methods described above. Otherwise, it may be possible to use a smaller, cheaper heatsink by taking account of the pulsed nature of the power dissipation. Figure 6.5 illustrates why this is so: the power device itself and the heatsink possess *thermal capacitance* (analogous to electrical capacitance). If the power dissipation is intermittent, the thermal capacitances are able to absorb heat during the time that power is being dissipated in the semiconductor device and release this heat through the thermal resistances during the remaining time. In reality, the thermal capacitances are *distributed* physically throughout the power device and heatsink and are not *lumped* as shown in the figure. This means that an analytical approach to heatsink calculations under pulse conditions is not feasible: instead we use empirical data provided by the device manufacturers from experiments and tests.

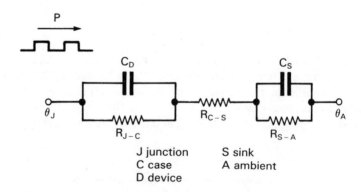

Fig. 6.5   Electrical analogue circuit for a power device on a heatsink (transient power dissipation).

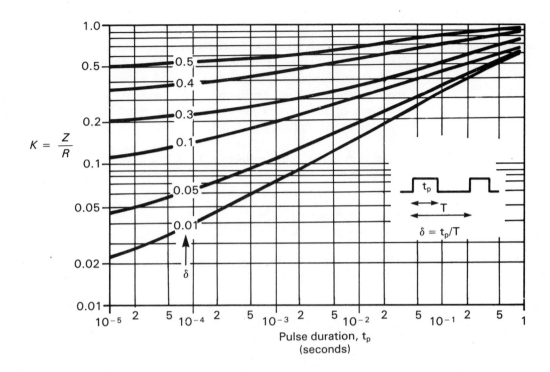

Fig. 6.6 Typical transient thermal impedance curves for a power semiconductor device.

Figure 6.6 shows the form of the data provided by the manufacturers of a power device operating under pulse conditions. The quantity δ is the duty cycle of the device, the ratio of on-time, t_p, to the period of the pulse waveform, T; K is the ratio of transient thermal impedance Z to steady-state thermal resistance R. With a pulsed power dissipation, eqn (6.5) becomes

$$\theta_J - \theta_A = P_{max}(KR_{J-C} + \delta R_{C-S} + \delta R_{S-A}) \qquad (6.6)$$

The thermal resistance from junction to case has been multiplied by the factor K, found from the manufacturer's data. Since K is less than 1, the effective *thermal impedance* from the junction to the case is lower, reflecting the fact that the mass of the device can absorb heat temporarily during the power pulses (in its thermal capacitance). The thermal resistances of the mica washer and the heatsink have simply been multiplied by the duty cycle of the pulse waveform, δ, to allow for the average power dissipation. (This is conservative: in fact a practical heatsink can have substantial thermal capacitance.)

If the actual operating power waveform of the device does not consist of rectangular pulses, an equivalent rectangular pulse waveform must be used to find δ and K. Clearly the areas under the actual and equivalent waveform pulses must be the same so that they contain equal energies. One can keep either the peak amplitudes equal or the pulse durations equal: the choice depends on the actual pulse waveform and is largely a matter of engineering judgement. (One could, of course, calculate the required heatsink thermal resistance for both cases and use the more conservative value in selecting a heatsink.)

Worked Example 6.2   A power transistor is used to switch a resistive load with a duty cycle of 30% and a pulse duration of 20 ms. When conducting the device dissipates 25 W. The thermal resistance, junction–case, is $1.5°C\ W^{-1}$ and a mica washer adds $0.3°C\ W^{-1}$. What is the maximum thermal resistance of the heatsink required for an ambient temperature of 50°C and a junction temperature of 150°C? Use the transient thermal impedance curves of Fig. 6.6.

Solution   From Fig. 6.6 $K = 0.4$ ($\delta = 0.3$, $t_p = 2 \times 10^{-2}$ s)
Using eqn (6.6)

$$150 - 50 = 25(0.4 \times 1.5 + 0.3 \times 0.3 + 0.3 \times R_{S-A})$$
$$R_{S-A} = 11°C\ W^{-1}$$

Exercise 6.3   If, in Worked Example 6.2, the device dissipation was assumed to be a *steady* 25 W (a conservative assumption), what would be the maximum thermal resistance of the heatsink? (*Answer*: $2.2°C\ W^{-1}$)

Forced cooling

The possibility of enhancing convective heat loss from a heatsink by forcing air across the surface of the heatsink has already been mentioned. Forced air cooling can also be used to remove heat from PCBs mounted in an equipment cabinet or rack, thereby reducing the operating temperature of components mounted on the PCBs and improving their long-term reliability. Figure 6.7 illustrates a typical application of forced air cooling. A small axial fan of about 100 mm diameter draws air through a dust filter and forces the air between a number of PCBs. The air picks up heat from the components on the PCBs and is exhausted through louvres or slots. Electric fans for this type of application are readily available in a range of sizes and operable from either a mains supply or a low-voltage d.c. supply. A volumetric air flow rate of around 100 m^3 $hour^{-1}$ can be produced by a fan of 100 mm diameter.

Volumetric flow rates may still be quoted in cubic feet per minute (CFM) by some manufacturers. 1 CFM $\cong$ 1.70 m^3 $hour^{-1}$.

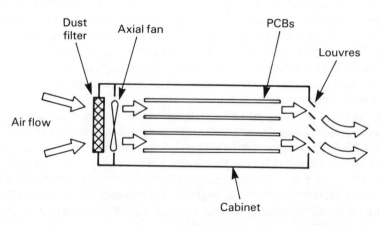

Fig. 6.7   A typical arrangement for force-cooling PCBs.

Derive an expression for the volumetric air flow rate required to limit the internal temperature of an enclosure to $\theta°C$ above ambient when the power dissipation inside the enclosure is P.

Solution When steady-state conditions have been attained the air flowing through the enclosure must remove P watts, or $3600 \times P$ joules hour^{-1}. If the specific heat capacity of air is c (J kg^{-1} °C^{-1}) then the mass flow rate needed is $3600P/c\theta$(kg hour^{-1}). Dividing by the density, ρ(kg m^{-3}) we obtain the volumetric flow rate required:

$$\frac{3600P}{\rho c\theta} \text{ (m}^3 \text{ hour}^{-1})$$

The air flow rate produced by a fan depends upon the resistance to air flow presented by filters, louvres and PCBs. Each resistance causes a pressure drop (analogous to voltage drop across an electrical resistance) and the total pressure drop must be overcome by the fan. Fan manufacturers state the performance of their products by means of a characteristic showing pressure difference against volumetric flow rate. The characteristic of a dust filter, or other air flow resistance, would be plotted in the same way, showing pressure *drop* versus flow rate. The flow rate achievable by a given fan with given air flow resistance characteristics for the filter and enclosure can be found by superimposing the two sets of characteristics as shown in Fig. 6.9. In practice, however, a fan can often be selected by calculating the required air flow rate as in Worked Example 6.3 and then applying empirical correction factors provided by filter and fan manufacturers to allow for air flow resistance.

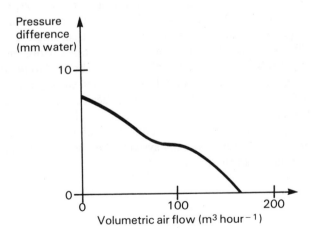

Fig. 6.8 Typical air flow characteristic for a small axial fan of about 100 mm diameter.

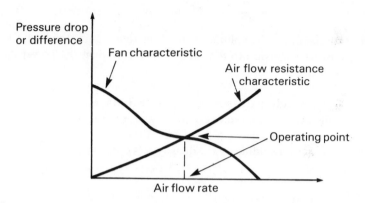

Fig. 6.9 Graphical determination of air flow rate from fan and air flow resistance characteristics.

Advanced Heat-removal Techniques

In many modern electronic engineering applications, the techniques so far described are sufficient to deal with heat-removal problems. Most printed circuit boards can be cooled by natural convection or gentle fan-assisted convection with no special design measures necessary on the PCB. Indeed some digital electronic systems implemented in CMOS logic consume so little power (and therefore dissipate so little heat) that they can be housed in a totally sealed enclosure. The reason that these systems present no special heat-removal problem can be regarded as a low *power density* (power dissipation per unit volume or per unit of PCB area). In applications where space is at a premium, such as on board an aircraft, designers may be forced to increase packing density (and therefore power density). The intro-

Surface-mount technology was described in Chapter 2.

duction of surface-mount technology, with its reduced component size, is also tending to incr̲ ̲e power densities.

The main problem with heat removal from a high-density system is the low thermal conductivity of PCB laminates. Multi-layer boards with internal power and ground planes have a higher thermal conductance than single- or double-sided boards, but still pose a problem in ultimate heat removal from the board. One possible solution to this problem is a metal heat ladder bonded to the PCB between the rows of ICs. Special provision is made at the board edge to remove heat from the heat ladder for conduction to the outside surface of the equipment.

In high-speed computers, propagation delays due to the finite speed of light become significant. This is discussed further in Chapter 7.

In very high-speed computers, components have to be placed physically close together, resulting in power densities so high that liquid (usually water) cooling has to be used. The physical design of such systems then becomes dominated by the coolant system rather than the PCB.

Summary

Heat is an unavoidable consequence of the flow of electric current and is therefore produced in all electronic equipment wherever electric current flows or circulates. Heat transfer can take place by three physical processes: conduction; convection; and radiation. Hot electronic components are less reliable than cooler ones.

Practical heat calculations use the notion of thermal resistance and an analogy with electrical resistance in which temperature difference replaces voltage and heat flow rate replaces electric current.

Power transistors and integrated circuits are kept cool by attaching them to heat-sinks. For steady-state power dissipation, the thermal analogue of Ohm's law can be used to select a heatsink or calculate the internal temperature or maximum allowable power dissipation of a power device. Under transient conditions, thermal capacitance can be included in the calculations to enable a smaller heatsink to be used.

Forced air flow can enhance heat loss from a heatsink or PCB.

In electronic systems with a high power density, special measures may be needed to remove heat.

Problems

6.1 Two transistors, each dissipating 15 W and having a thermal resistance (junction–case) of $1.5°C\ W^{-1}$, are to be mounted together on the same heat-sink. The junction temperatures are to be limited to 60°C above ambient temperature. Assume the thermal resistance of mica washers and thermal grease to be $0.4°C\ W^{-1}$. What is the maximum allowable thermal resistance of the heatsink?   0.1

6.2 A designer wishes to use a small transistor dissipating 1.2 W. The manufacturer's data sheet does not state a thermal resistance value but gives the power-dissipation derating curve shown in Fig. 6.3. A heat dissipator with a thermal resistance of $48°C\ W^{-1}$ is available to fit the transistor. What is (a) the maximum safe dissipation with the dissipator; and (b) the case temperature at a power dissipation of 1.2 W? Assume an ambient temperature of 25°C.

$R_\theta = \frac{1}{20} = 0.05$

$P = 1.2\text{W}$

6.3 A cabinet containing circuitry dissipating 150 W is to be cooled by a ventilating fan. The ambient temperature inside the cabinet is to be limited to 50°C for an air inlet temperature of 30°C. What is the minimum air flow rate required? The specific heat capacity of air at 30°C is about $1000\ J\ kg^{-1}\ °C^{-1}$ and the density of air at 30°C and normal atmospheric pressure is about $1.3\ kg\ m^{-3}$.   20.76

7 Parasitic Electrical and Electromagnetic Effects

Objectives

☐ To introduce the problem of parasitic circuit elements and the limitations of lumped-parameter circuit models.

☐ To review electromagnetic induction effects.

☐ To introduce the problem of electromagnetic interference and the subject of electromagnetic compatibility.

☐ To present studies of common problems in real circuits and examples of good practice in circuit and system design.

The behaviour of many electronic circuits can be influenced by parasitic electrical and electromagnetic effects which are incidental to the intended properties of the circuit. Troublesome symptoms in analogue systems include *crosstalk* or unwanted coupling of a signal from its intended pathway to some other pathway; *instability* or spurious oscillation; *pickup* of unwanted signals especially at mains frequencies and *microphony* or sensitivity to mechanical disturbance or vibration. Digital systems can be prone to data errors and spurious states caused by crosstalk and switching transients. Both types of system can be susceptible to electrical or electromagnetic interference (EMI), which is often caused by other electrical or electronic equipment, and can themselves be sources of such interference.

Many of these problems can be controlled, often by the application of quite simple design measures and good practice. In order to understand how these design measures work we must first examine mechanisms which allow electromagnetic energy to stray from its intended path.

Parasitic Circuit Elements

Parasitic circuit elements are not restricted to passive devices. In integrated circuits, parasitic active elements such as transistors are often inherent in a particular IC process.

Any electronic circuit contains parasitic or 'stray' circuit elements. Some of these are inherent in the circuit components, as discussed in Chapter 5, while others are properties of the physical layout and surroundings of the circuit. Sometimes the parasitic elements have very little effect on circuit performance and can be ignored. In other cases, however, the ultimate level of performance achievable from a circuit can be determined by parasitic effects. The three types of passive parasitic circuit elements will now be considered.

Important exceptions include the resistance of power supply rails and internal series resistance in some integrated circuits.

Parasitic resistance is not often a problem since the resistance of a wire or PCB track is usually negligible compared to the impedances of a circuit, and the insulation resistance between wires or tracks is so much higher than the circuit impedance, that voltage drops and leakage currents have negligible effect. Nevertheless, it is as well to remember that voltage drops and leakage currents do exist and that insulation can break down at sufficiently high voltages.

Parasitic reactance is a much more common problem in electronic circuits. It can be understood in terms of parasitic capacitance and inductance, or in terms of the equivalent electric and magnetic fields generated by a circuit. Both points of view are valid and useful, but as we shall see, the field concept is more general and gives a better understanding of control measures used to reduce the magnitude of the reactances. A *parasitic capacitance* exists between any pair of conductors and can couple energy from one conductor to the other. Similarly *parasitic inductance* can exist either as self-inductance, which can be important in power supply circuits, particularly in digital systems, or because any pair of circuits has a mutual inductance, which can couple energy from one circuit to another.

Strictly, only a closed circuit can have self-inductance since there must always be a return path for current. Later in this chapter it is explained that the self-inductance of a circuit is reduced if the out and return conductors are kept close together.

Parasitic Capacitance

The capacitance between two conductors can be defined as $C = Q/V$, where the presence of equal and opposite charges of magnitude Q on the conductors results in a potential difference V between them. An electric field exists between any two oppositely charged conductors and the line integral of this field $\int\mathbf{E.dl}$ along any path between the conductors is equal to the potential difference V. We can associate capacitance therefore, with the electric field between conductors caused by the presence of charge on the conductors. The capacitance between two conductors can be modified by altering the electric field distribution in some way. Normally we wish to reduce parasitic capacitances and several ways of doing so are discussed later.

Before going any further, it is useful to have some feel for the magnitude of a parasitic capacitance. We could consider two wires and calculate the capacitance between them, but there is a much easier approach. We can look in an electronic components catalogue to find the value of the smallest capacitor we can buy. This turns out to be about 1 pF. Suppose we were to create a capacitor of this value using the two sides of a PCB as the plates. What area of copper would be needed, neglecting fringing effects at the edges of the plates?

Both Compton and Carter discuss calculation of stray capacitance.

A parallel plate capacitor has a capacitance $C = \epsilon_0\epsilon_r A/d$. What area, A, is required to form a 1 pF capacitor if d, the plate separation, is 1.6 mm (a typical PCB thickness) and ϵ_r for PCB laminate is about 6?

Worked Example 7.1

Solution To a good approximation for this purpose ϵ_0 is 9 pF m^{-1}. The value of $C/\epsilon_0\epsilon_r$ is thus $\frac{1}{54}$ and A is $1.6 \times 10^{-3}/54$ or about 30 mm². If the plates are square they will be about 5 mm × 5 mm.

We can see from this why capacitors smaller than 1 pF are not commonly available and also that quite small areas of copper on a PCB have a capacitance of this order. Looking again in a catalogue we can find that an RG58C/U coaxial cable has a capacitance of 100 pF m^{-1} or 1 pF cm^{-1}; that a non-screened twisted pair cable has a capacitance of 56 pF m^{-1} and that a 300 Ω balanced feeder cable which has two straight conductors spaced about 10 mm apart by a plastic web has a capacitance of 13 pF m^{-1}. We could therefore make a reasonable estimate of the capacitance between two PCB tracks or wires a few millimetres apart at, say, 10–20 pF m^{-1}.

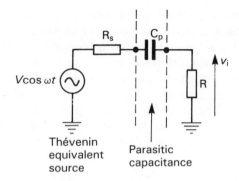

Fig. 7.1 Parasitic capacitive coupling between a source and a load.

Because the impedance of a capacitor is inversely proportional to frequency, parasitic capacitances have a greater effect in high-frequency circuits and where high-frequency interference is present. Parasitic capacitances tend to couple high-frequency signals and interference into high impedance circuits as shown by the equivalent circuit in Fig. 7.1. The Thévenin equivalent source network represents a circuit node where a high-frequency signal is present, coupled through a parasitic capacitance, C_p, to a circuit node with a high impedance to ground, represented by R. v_i is the resulting interference or unwanted signal present across R. This circuit is a potential divider, so that v_i is given by

The sinusoidal voltage $V \cos \omega t$ is represented by the real part of the complex exponential $V \exp (j\omega t)$. Equation (7.1) is obtained by taking the modulus of the complex quantity v_i.

$$v_i = \frac{R}{R + R_s + 1/j\omega C_p} \cdot V \exp (j\omega t)$$

and the magnitude of v_i is

$$|v_i| = \frac{VR}{\sqrt{((R + R_s)^2 + \cdot(1/\omega^2 C_p^2))}} \qquad (7.1)$$

We want the interference voltage v_i to be as small as possible, which means that the denominator in eqn (7.1) must be much greater than R.

Worked Example 7.2 Assume that R in Fig. 7.1 represents the input impedance of a moderately high-impedance amplifier, say $R = 100$ kΩ. Let C_p take a typical 'stray' value of 0.1 pF. If the Thévenin source network represents an interference source such as a nearby power-supply track carrying high-frequency interference, then the value of R_s may be neglected. At what frequency would the magnitude of v_i be 60 dB less than V? (60 dB represents a voltage ratio of 1000.)

Solution As $R_s \ll R$ we may approximate eqn (7.1) to give

$$\frac{|v_i|}{V} = \frac{R}{\sqrt{(R^2 + (1/\omega^2 C_p^2))}} = 10^{-3}$$

or

$$\frac{1}{\sqrt{(1 + (1/\omega^2 C_p^2 R^2))}} = 10^{-3}$$

which gives approximately $\omega C_p R = 10^{-3}$; hence $\omega = 10^{-3}/RC_p$. Since $RC_p = 10^5 \times 10^{-13} = 10^{-8}$, $\omega = 10^5$ rad/s or 16 kHz.

Notice that the frequency is in the audio band. Such effects are not only important at high frequencies.

We can conclude from this example that any high-impedance node in a circuit is highly prone to pick up interference from a nearby source by capacitive coupling. We must therefore be careful when designing the physical layout of the circuit to minimize the parasitic capacitance between any likely interference sources and any high-impedance nodes. We can do this in several ways, the most obvious of which is to keep interference sources well away from the sensitive node, since increased separation reduces the parasitic capacitance. There are two much more effective ways of reducing a parasitic capacitance between two nodes, shown in Fig. 7.2. One is to interpose an earthed conducting screen between the two nodes, the other is to place the two nodes close to an earthed surface or *ground plane*. The screen or ground plane alters the electric field distribution between the conductors and reduces the field strength near to one conductor due to charge on the other. It also increases the capacitance to ground of each conductor. This is an inevitable price to be paid for the reduction in mutual capacitance.

Carter discusses the calculation of capacitances using finite element methods on a digital computer. The electric field distribution can also be mapped by these methods.

The principle illustrated in Fig. 7.2(a) can be extended so that the screen totally surrounds one of the conductors. This is known as a *Faraday cage* and it relies on the well-known result in electrostatics that there can be no electric field inside a conductor, and thus no electric field inside a region enclosed by conductor due to a field outside the enclosure. Note carefully the word *electrostatics*: we are talking here only about static electric fields. In practical circuits charges move and generate *electromagnetic* fields. We can begin to look at the problems caused by these by discussing inductance.

The Faraday cage is widely used in screened rooms for electromagnetic testing (discussed in Chapter 9). It is also applied when surrounding the front-end of radio receivers, where weak r.f. signals are being amplified, to prevent pickup from powerful local sources within the receiver.

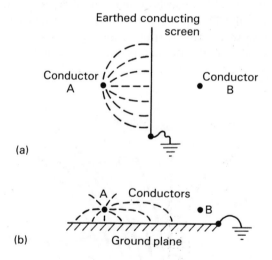

(a)

(b)

Fig. 7.2 Two methods of reducing the capacitance betwen a pair of conductors: (a) an earthed screen; (b) a ground plane.

Parasitic Inductance

Parasitic inductance occurs in two forms. One is the *self-inductance* of a circuit such as a power-supply rail, load and return; the other is the *mutual inductance* between two circuits which can couple unwanted energy from one circuit to another. As we shall see, self-inductance and mutual inductance are closely related, so that measures to reduce one can also reduce the other.

Figure 7.3 shows two circuits where the magnetic flux generated by loop 1 links loop 2 and vice versa. Self- and mutual inductance are associated with the magnetic field around a conductor generated by a current. The mutual inductance, M, between two loops can be expressed in terms of the self-inductances of the loops L_1 and L_2 as $M = k\sqrt{(L_1 L_2)}$, where k is a dimensionless factor between 0 and 1 which depends on the proportion of the flux generated by one loop which links the other loop. We cannot easily get a feel for the magnitude of 'stray' inductances as we did for stray capacitances, but as we shall see later we can find the self-inductance per unit length of a coaxial or twisted-pair cable. For the RG58C/U cable this turns out to be 250 nH m^{-1} and for the 300 Ω balanced feeder cable with two conductors spaced about 10 mm apart, 1 μH m^{-1}. We can say therefore that the mutual inductance between two pairs of wires running parallel and a few millimetres apart is of the order of 100 nH m^{-1} to 1 μH m^{-1}. To reduce the mutual inductance between circuits and therefore minimize electromagnetic interference problems we need to reduce the flux linkage between the circuits. This can be done in several ways. In some situations the flux generated by loop 1 is outside our control. It might be the 50 Hz flux generated by nearby mains wiring. The flux passing through loop 2, however, can be reduced by reducing the area of loop 2. If the wires are twisted together, not only is the area of the loop reduced, but the direction of magnetic flux lines passing between the two wires alternates with each twist. The effect is as if the flux lines had been made to alternate, cancelling out over the length of the circuit. The flux generated by loop 1 can be reduced by the same method. Wiring to or from a transformer or power supply should therefore be designed so that both conductors are close together throughout their length.

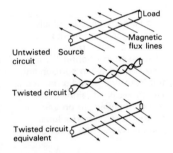

Untwisted circuit

Twisted circuit

Twisted circuit equivalent

As we shall see later, the *self-inductance* of a circuit is also reduced by running the conductors close together and this is also desirable.

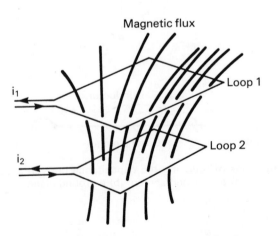

Fig. 7.3   Magnetic flux linking two circuit loops.

There are three other methods of reducing inductive coupling between circuits, two of which are analogous to those used to reduce capacitive coupling whereas the other has no capacitive analogue.

As with capacitive coupling, increasing the physical separation between two circuits reduces inductive coupling. Quite often, when designing an electronic system, it can be quite easy to route power supply wires carrying heavy current away from sensitive circuits. Secondly in the same way that an electric screen could be used to reduce coupling by an electric field, a magnetic shield can be used to reduce coupling by a magnetic field. In the electric case, the screen must be of high conductivity, which is easily achieved with copper. In the magnetic case, however, high permeability is needed. A magnetic shielding alloy, such as *mumetal* can be used, but it is not as effective against a magnetic field as copper is against an electric field. Magnetic shields are also less effective at very low frequencies.

The third technique mentioned above, which has no analogue in capacitive coupling, is to position loop 2 relative to loop 1 so that the lines of flux generated by loop 1 are parallel to the plane of loop 2.

One final step that we might be able to take to reduce inductive coupling is unconnected with the actual mutual inductance between two circuits. The flux generated by a circuit loop is proportional to the current flowing in the loop. Therefore reducing the current can reduce coupling problems. As an example of the application of this idea, if a number of PCBs are connected via a backplane, the boards taking the greatest current could be positioned at the end of the backplane nearest to the power-supply feed, so that the loop of greatest area does not carry the full current.

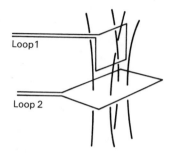

Magnetic flux generated by loop 1.

Distributed-parameter Circuits

So far in discussing parasitic effects in electronic circuits a conventional circuit approach has been used, supplemented in the case of capacitive and inductive coupling between circuits, by some consideration of electric and magnetic fields. Circuit models are only an approximation to reality and we might ask under what conditions a circuit model ceases to be valid and what then happens? To answer these questions it is helpful to look at the behaviour of *transmission lines*, because it is here that our conventional circuit models begin to fail us. The example presented below can easily be demonstrated in the laboratory and gives a good intuitive feel for the subject.

Figure 7.4 shows a circuit in which a voltage source generates a short pulse of amplitude A and duration T. A load resistance R is connected to the source. Circuit theory tells us that the voltage v_R across the resistor will be identical to the voltage v_S at the source and that the current i will be a pulse of amplitude A/R and of the same duration T as the voltage pulse. In carrying out this simple analysis we have neglected, among other things, the parameter d, the distance from the voltage source v_S to the load resistance R, shown in the figure. To see why d can be important imagine T to be 10 ns and d to be 10 m. How fast does the pulse travel along the wires from the voltage source to the load resistor? Well, it cannot travel faster than light, so we will take this to be the speed of the pulse. The speed of light is, to a good approximation, 3×10^8 m s^{-1} or, in more suitable units for electronic engineering, 300 mm ns^{-1}. The leading edge of the 10 ns pulse will therefore travel

In fact the speed of an actual pulse can be as low as 100 mm ns^{-1} or $\frac{1}{3}$ of the speed of light.

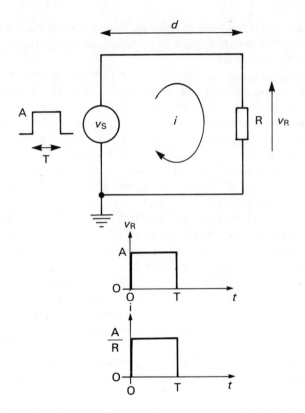

Fig. 7.4 A pulse circuit to illustrate the importance of circuit dimensions.

3 m before the trailing edge leaves the voltage source and the leading edge of the pulse will not reach R until 33 ns after leaving the voltage source. It follows therefore that R can have no effect on the current flowing from the voltage source and our conventional circuit model fails. The reason is that the duration of the signal, T, is less than the size of the circuit, d, divided by the speed of light, c.

$$T < \frac{d}{c} \tag{7.2}$$

Under these conditions, *lumped-parameter* circuit models in which circuit properties are considered to be localized in components are not valid, and we can no longer neglect the interconnecting wiring. Instead, we must consider the wiring as a *transmission line* and design the wiring to have the necessary transmission-line parameters for our application. The electrical energy travelling along a line can be thought of as an electromagnetic wave *guided* by the conductors.

Carter, Chapter 7, contains a much more thorough treatment of transmission lines than is presented here.

A transmission line has a *characteristic impedance Z_0* which depends on the geometry and dimensions of the line conductors and the permittivity of the medium or dielectric between the conductors. Z_0 is a dynamic impedance (it cannot be measured with a multimeter) and is independent of the length of the line. The input impedance of a transmission line is Z_0 and the output impedance or Thévenin equivalent source impedance is also Z_0. In the example shown by Fig. 7.4 the current flowing from the voltage source is determined by the characteristic impedance of the transmission line linking the source to the load resistance and not by the value of the load resistance R. As we shall see, the behaviour of the pulse on arrival at the load depends on the value of R and the value of Z_0 and some interesting effects can occur.

Z_0 is related to the *distributed* parameters of the transmission line

$$Z_0 = \sqrt{\left(\frac{R + j\omega L}{G + j\omega C}\right)} \qquad (7.3)$$

Kraus and Carver give a derivation of eqn (7.3).

where R is the series resistance per unit length (Ω m^{-1}), L is the series inductance (H m^{-1}), G is the shunt conductance (Ω^{-1} m^{-1}) and C is the shunt capacitance (F m^{-1}). On an ideal or *lossless* line the series resistance and shunt conductance of the line are zero and eqn (7.3) reduces to $Z_0 = \sqrt{(L/C)}$. A practical transmission line has some resistance and conductance that absorb energy from any wave or pulse propagating on the line. The loss is quantified by an attenuation coefficient α which is typically one or two dB per 100 m, increasing with frequency.

The speed with which a wave or pulse travels down the line is determined by the relative permittivity of the medium or dielectric between the conductors. The propagation velocity is given by

$$v = \frac{1}{\sqrt{(LC)}} = \frac{1}{\sqrt{(\epsilon_0 \epsilon_r \mu_0)}} = \frac{c}{\sqrt{\epsilon_r}} \qquad (7.4)$$

where L and C are the distributed inductance and capacitance per unit length, ϵ_0 is the permittivity of free space, ϵ_r is the relative permittivity of the dielectric, μ_0 is the permeability of free space and c is the speed of light.

An RG58C/U coaxial cable has a characteristic impedance of 50 Ω and a shunt capacitance of 100 pF m^{-1}. What is (a) the series inductance, (b) the propagation velocity and (c) the relative permittivity of the dielectric?
(*Answers*: (a) 250 nH m^{-1} (b) 200 mm ns^{-1} (c) 2.25.)

Exercise 7.1

What is the approximate propagation velocity on a PCB track if ϵ_r = 6 for PCB laminate?
(*Answer*: 120 mm ns^{-1}. In practice the electric field extends outside the dielectric and the true velocity is a little higher.)

Exercise 7.2

If a length of transmission line with characteristic impedance Z_0 is *terminated* with an impedance Z, what happens to a pulse travelling along the line on arrival at the termination? The answer is not only interesting but is of great significance in the design of high-speed digital systems such as computers. Figure 7.5 shows three special cases. In (a) the end of the line is left open circuit, and since no energy can be absorbed by an open circuit, the incident pulse is *reflected* and travels back along the line to the source. During the time that the pulse is being reflected, the incident and reflected parts of the pulse overlap so that the voltage present at the end of the line is twice the pulse amplitude. This effect is known as *voltage doubling*. In (b) the line is terminated with a resistance equal to the characteristic impedance of the line. There is no reflection and this is thus a desirable situation if the line is carrying a signal from one part of a computer to another, since reflections could cause errors. The absence of reflection can be explained by imagining the terminating resistance to be replaced by a length of the same line extending to

There is also the possibility that the energy in the pulse could radiate from the open end of the line as an electromagnetic wave. In practice, a purpose-designed *antenna* is needed if any significant energy is to be radiated. A perfectly *matched* antenna would have an impedance of Z_0 and would radiate all the energy into free space.

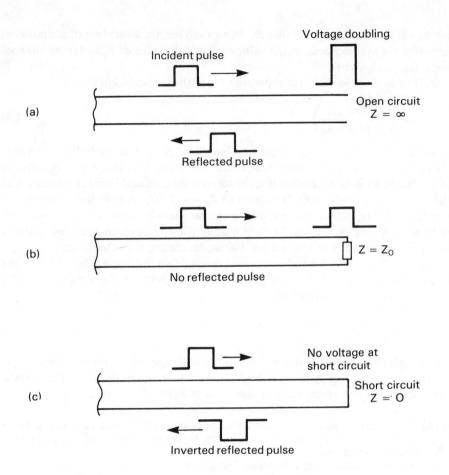

Fig. 7.5 Transmission line termination effects: (a) voltage doubling at an open circuit; (b) matched termination; (c) pulse inversion at a short circuit.

infinity. The input impedance of this semi-infinite line is Z_0. Once a pulse has been launched down an infinite length of line we shall never see it again. An equivalent argument is that there is no reflection from a joint between two identical lines and therefore there is no reflection from a joint between a line and a resistance equal to Z_0. In (c) the line is terminated with a short-circuit. As in case (a), no energy can be absorbed at the termination, and since there can be no voltage at a short circuit, the reflected pulse is inverted.

In the general case of a transmission line of characteristic impedance Z_0 terminated with an impedance Z, the amplitudes of the incident and reflected pulses are given by the equation

$$\frac{V_r}{V_i} = \frac{Z - Z_0}{Z + Z_0} \tag{7.5}$$

Exercise 7.3   This is a practical exercise. Set up an oscilloscope of at least 50 MHz bandwidth, a pulse generator and a length of 50 Ω coaxial cable as shown in the margin. The pulse generator must have a 50 Ω output so that any reflected pulses are absorbed when they arrive back at the pulse generator. With 30 m of cable the pulses will need to be of about 100 ns duration.

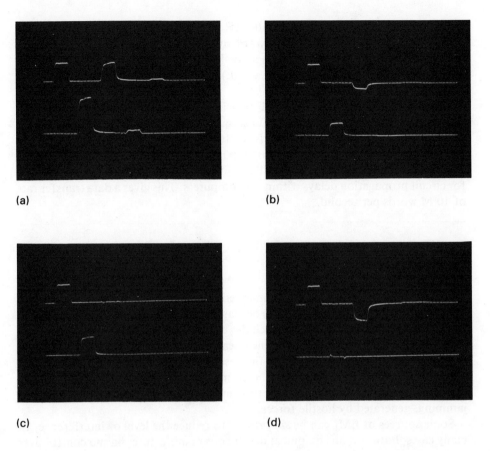

(a)

(b)

(c)

(d)

Fig. 7.6 Pulse reflections on 28 m of 50 Ω coaxial line. Pulse width 100 ns. Top trace: transmitting end; bottom trace; termination end. (a) Open circuit termination; (b) 27 Ω termination; (c) 50 Ω termination; (d) short-circuit termination.

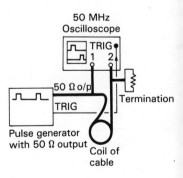

Experiment with a variety of terminations including open and short circuits, 25 Ω, 50 Ω and 100 Ω and verify the behaviour discussed above. (The input impedance of the oscilloscope can be considered infinite if of 1 MΩ or more. Do not use an oscilloscope with a 50 Ω input.)

Figure 7.6 shows some results obtained by the author using 28 m of cable and a 100 ns pulse as described in Exercise 7.3. Four cases are shown. The top trace in each case shows the voltage measured at the pulse generator, the bottom trace shows the voltage measured at the far end of the cable. The pulse generator is not perfectly *matched* to the cable so that a secondary reflection occurs when the pulse arrives back at the pulse generator.

Pulse propagation on transmission lines is very important in the design of high-speed digital systems.

Consider a parallel data link between two computers 5 m apart, made from ribbon cable with a characteristic impedance of 105 Ω and a capacitance of 50 pF m^{-1} between adjacent conductors. If handshaked data transmission is used, what is the

Worked Example 7.3

maximum possible data transfer rate? (Handshaking means that the receiving computer acknowledges each data word by sending a signal back to the transmitting computer. Only when the acknowledgement is received does the transmitting computer send the next data word.)

Solution   The capacitance is 100 pF m^{-1} if alternate conductors are grounded. Assuming the cable to be lossless and using $Z_0 = \sqrt{(L/C)}$ and eqn (7.4), the speed of propagation v is $1/Z_0 C$ which is $1/(105.100 \times 10^{-12})$ or 95 mm ns^{-1}. The propagation time over 5 m is thus 5000/95 or 52 ns and allowing for the acknowledgement each data word will require about 100 ns, making no allowance for circuit propagation delays within the computers. This gives a data transfer rate of 10 M words per second.

Electromagnetic Interference

Any unwanted electrical or electromagnetic energy which disturbs the normal operation of an electronic system can be classed as *electromagnetic interference* or EMI. Sources of EMI in industry include: electric motors, electric furnaces, arc-welding equipment and radio-frequency heating and drying systems. In the home and the office, EMI can be caused by fluorescent lights, dimmer switches, personal computers, thermostats, hairdryers, vacuum cleaners and power tools. Military systems may also be exposed to radar transmissions and deliberate interference, or jamming, generated by hostile forces.

Some sources of EMI can be *suppressed* to reduce the level of interference. In many cases, however, an equipment designer or manufacturer has no control over the environment in which his product will be required to operate and the design must be capable of working in the presence of interference. Badly designed products can also be sources of interference, releasing unintended electromagnetic energy into the surrounding environment and causing an EMI problem elsewhere.

The study of EMI problems involving the effects of one system upon another, and within a complex system, is known as *electromagnetic compatibility* or EMC, and was first developed as a distinct discipline by electronics engineers working on military systems. EMC problems can occur in military aircraft, ships and fighting vehicles because of the close proximity of many separate, sophisticated electronic sub-systems for communication, navigation, target detection, range-finding and weapon control. The increasing use of electronics in civilian applications is making EMC an important aspect of all electronic product design. International standards exist to define the acceptable levels of EMI generated by an electronic product.

EMI Mechanisms

There are two main ways in which EMI can enter (or leave) an electronic system: *conduction* along wiring and cabling; and *radiation* through free space. Radiated EMI can be further divided into *induction fields* from nearby sources and *plane-wave fields* from far sources. Transmission line effects were introduced in the previous section, but so far the generation and reception of radiated EMI has not been discussed. Devices intended to launch or receive electromagnetic waves into or from free space are known as *antennas*. Circuits can exhibit parasitic antenna effects and may generate not only local electric and magnetic induction fields which can couple into other nearby circuits but also electromagnetic waves which can propagate farther away and cause interference to distant systems. Parasitic antennas may also receive interference from distance sources.

Antennas are discussed by Kraus and Carver.

Far from an antenna, in terms of signal wavelength, the electromagnetic waves generated have a constant ratio of electric field strength to magnetic field strength of 377 Ω. This is known as the intrinsic impedance of free space. From electromagnetic theory it can be shown to be given by $\sqrt{(\mu_0/\epsilon_0)}$ where μ_0 and ϵ_0 are the permeability and permittivity of free space respectively. Close to an antenna, in the *near field* the ratio of electric field strength to magnetic field strength or the *field impedance* can be less than or greater than 377 Ω depending on the type of antenna, or circuit, generating the field. High-impedance fields, where the electric field component dominates, are generated by high-impedance circuits operating at high voltages and low currents. Low-impedance fields with a strong magnetic component are generated by high-current circuits with low series impedance and low voltage drops.

These statements should be compared with those made earlier in the discussion of parasitic capacitance and mutual inductance. They are simply another way of looking at the same phenomenon.

Figure 7.7 depicts the main routes by which EMI can enter an electronic system. All of these routes are reversible and EMI generated by the system can leave along the same pathways to cause trouble elsewhere. It is important to realize that interference conducting along, say mains wiring, may radiate once inside the equipment and that radiated EMI may couple into internal wiring and propagate by conduction from one sub-system to another.

We are now in a position to study techniques to control or reduce EMI problems, firstly from the viewpoint of susceptibility to external EMI and secondly from the EMC viewpoint in reducing EMI emissions from a system. As we shall see, there are techniques which are helpful from both points of view.

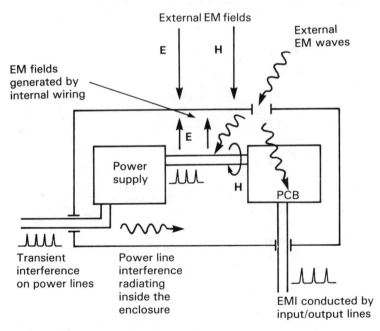

Fig. 7.7 Routes by which EMI can enter or leave a system.

When we are designing an electronic product or system we may not have a detailed knowledge about the EMI environment in which our design is to work. This can make reduction of EMI susceptibility a somewhat empirical activity. In some designs there will be specific EMI sources nearby operating on known frequencies and at known power levels. Protection measures against interference can then be designed with greater confidence using EMC manuals and calculation to analyse the performance of a proposed protective measure. There is not sufficient space in this book to discuss such matters and the material that follows is intended to outline general principles only.

A more thorough treatment is given by Keiser who discusses EMC in detail and gives references to specialist literature.

Before discussing techniques for preventing EMI entering a system it is worth examining the possibility of designing a system to be inherently immune to EMI. One way in which this can be done is to limit the bandwidth of the system to just

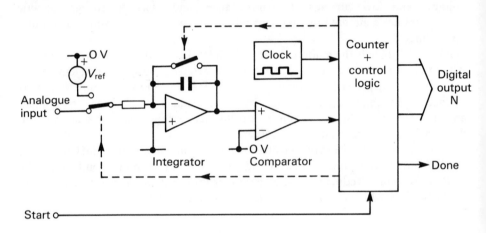

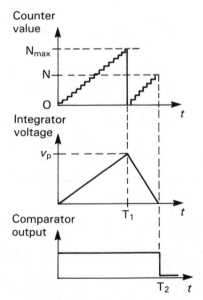

Fig. 7.8   A dual-slope integrator ADC.

that range of frequencies necessary for the system to do its job. A thermocouple amplifier, for example, can be limited to a response time of the same order as that of a thermocouple, say 1 s. The frequency response of the amplifier can cut off therefore at less than 10 Hz, so that sensitivity to 50 Hz and r.f. interference will be reduced. Another possibility for reduction of mains frequency interference can be applied in the design of an averaging measurement system. If the averaging time is designed to be a multiple of the mains period, any interference at the mains frequency or its harmonics will average to zero over the measurement time. This technique can be used with a dual-slope integrator analogue-to-digital converter as shown in Fig. 7.8. This type of ADC operates in two stages. Initially, the integrator is held at zero by the switch across the integrator capacitor and the counter is held at zero. When the ADC is started, the analogue input is connected to the integrator which ramps up as the counter increments towards its maximum value N_{max}. The first stage of the conversion ends at time T_1 as the counter overflows. During the second stage of the conversion the integrator is connected to the negative reference voltage V_{ref} and ramps down towards zero. At time T_2 the comparator switches and the counter is stopped. The final value in the counter, N, is accurately proportional to T_2, which is in turn proportional to the *mean value* of the analogue input during the first stage of the conversion. Since T_1 is dependent only on the counter length and the clock frequency, a value which is an integral multiple of the mains period can be chosen. Any mains hum present in the analogue input signal is thus averaged out during the first stage of the conversion.

Any mains hum present in the dual-slope ADC itself is not of course averaged out.

Grounding

In circuit design we normally take for granted the concept of an *earth*, *ground* or 0 V reference potential. This concept is worth examining in a little detail, as a practical ground will depart from the ideal. Consider a signal source connected to a load as shown in Fig. 7.9. The source might represent a transducer such as a microphone, producing a small voltage, say 10 mV, while the load might represent the input impedance of an amplifier. There are three potential differences shown in the figure, of which v_G would be zero in a circuit with an ideal ground, leaving only the transducer e.m.f. and the voltage at the amplifier input, which would be equal by application of Kirchhoff's voltage law. In reality, of course, the ground and signal connections have some non-zero impedance, and a potential difference can

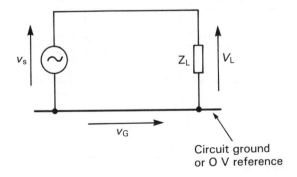

Fig. 7.9 A signal source connected to a load.

exist along their lengths if a current flows. Ground currents can be caused by external electromagnetic induction effects and by power-supply return currents. Both can be controlled by careful layout and design of a grounding system, or by avoiding grounding altogether and using a *floating ground*, which is simply a 0 V reference which is electrically isolated from mains or other earths.

Single-point grounding is an ideal arrangement, in which all ground connections within a system are made at a single point, so that from any circuit in the system there is only one path to earth. This avoids the creation of *ground loops* in which a closed loop of ground conductor exists. A ground loop can be coupled by induction fields, especially magnetic fields from nearby mains transformers, causing circulating currents and spurious voltages in the ground loop. Figure 7.10 shows a typical situation in which a ground loop exists and where the ideal of single-point grounding cannot be attained. Two electronic instruments are connected to mains earth by their mains supply leads, and the chassis or cabinet of each instrument is earthed for safety reasons as well as to shield against EMI. A signal is passed from one instrument to the other via a screened cable, and the cable screen is connected to the cabinets of both instruments, forming a ground loop as shown. Circulating currents in the ground loop may cause a potential difference along the length of the cable which *adds* to the signal voltage measured by instrument 2. There are various solutions to this problem depending on the frequency range and sensitivity of the instruments. One possibility is to separate the signal grounds within the instruments and the screen of the cable from mains earth with a resistance so that the impedance of the signal ground loop is increased. The obvious possibility of breaking the ground loop at some point cannot normally be used: both instrument cabinets must be earthed for safety reasons and if the cable screen is disconnected at one end, the signal return current must follow the mains earth path, which has a large self-inductance because of the loop area. A second possibility is to provide a high-quality low-impedance ground connection between the two instruments using a heavy copper braid *strap*.

At higher frequencies *multi-point* grounding must be used, because when the length of a ground lead approaches a quarter of the signal wavelength, the lead no longer provides a low-impedance path to ground. Ground leads must therefore be

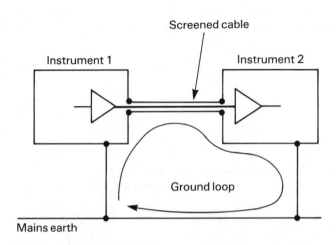

Fig. 7.10   The ground loop problem.

kept short and connected directly to a ground plane at the nearest available point. The multi-point ground must be of low impedance and carefully designed to maintain a low impedance over the operating life of the system. Materials must be chosen, for example, so that corrosion does not impair grounding effectiveness.

Power-supply Coupling

A signal in one circuit can constitute interference if it gets into a neighbouring circuit. Inter-circuit coupling can occur via the power supply circuit where two circuits share a power supply. Figure 7.11 shows how power-supply coupling can occur. In (a) two amplifiers, A1 and A2, share a common power supply such that A1 is downstream of A2. Z_s and Z_g are the impedances of the power supply and ground return conductors between the power supply and A2. I_2 is the current drawn by A2 and will have a steady component due to internal bias currents and a fluctuating component due to the signal being amplified by A2. As shown, the supply voltage V_1 at A1 is reduced by the voltage drop across Z_s and Z_g, so that a signal component from A2 can enter A1 via the power supply leads. Figure 7.11(b) shows a power distribution arrangement designed to overcome this problem. A2 is supplied by separate leads running back to the power supply, or at least to a point

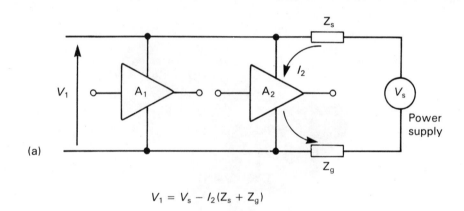

$$V_1 = V_s - I_2(Z_s + Z_g)$$

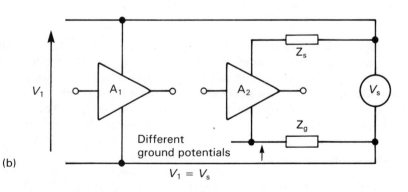

$$V_1 = V_s$$

Fig. 7.11   Power-supply coupling.

115

where a low-impedance path exists back to the power supply. The voltage drops across Z_s and Z_g now no longer affect the supply to A1, but there is now a possible problem with differing ground potentials at A1 and A2, and we could be back to the problem of providing a low-impedance ground return to reduce Z_g. Practical engineering problems are very often like this!

Filtering

Interference *conducted* along mains and other cables and internal wiring can be reflected or attenuated by *filters*, which allow wanted frequencies to pass while blocking unwanted frequencies. *Reflective* filters contain reactive components only and block conducted EMI by reflecting unwanted energy back along the wiring, in a similar manner to unmatched transmission-line terminations. *Lossy* filters, on the other hand, also have resistive elements (not necessarily in the form of lumped resistors) to absorb unwanted energy and dissipate it as heat.

Conducted interference is likely at frequencies of up to 30 MHz. Above this frequency interference tends to radiate from conductors rather than propagating as a guided wave. High-frequency interference does not radiate significantly from a purpose-designed high-frequency cable such as coaxial cable, so that interference above 30 MHz can be conducted well by such a cable.

Fig. 7.12   A commercial mains inlet filter with integral BS 4491 mains inlet connector. (Courtesy Schaffner EMC Ltd)

Mains filters with integral BS 4491 mains connectors of the type shown in Fig. 7.12 can be used to filter mains-borne interference. The construction of this type of filter means that there is no unfiltered mains wiring inside the equipment enclosure, eliminating radiated interference problems from the mains wiring. A common circuit for a single-phase mains filter is shown in Fig. 7.13. The operation of this circuit can be considered as a deliberate transmission line mismatch, or in terms of the conventional circuit model. The inductors, or *chokes*, L_1 and L_2 are wound on a common toroid so that the magnetic fluxes generated by normal a.c. currents passing through the filter cancel out and ensure that the core does not saturate. This means that the inductors present an unavoidable low impedance to *differential-mode* interference where the line and neutral conductors carry equal

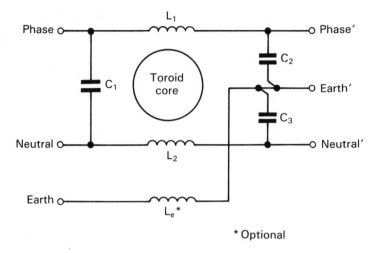

* Optional

Fig. 7.13 Typical equivalent circuit of a mains filter.

and opposite interference currents. *Common-mode* interference, however, is blocked by the high impedance of the two chokes to currents flowing equally in the two supply lines. The filter shown in Fig. 7.12, for example, gives an attenuation or insertion loss of only 10 dB at 150 kHz to a difference signal, but over 40 dB at the same frequency to a common-mode signal, when tested using a 50 Ω source and a 50 Ω load. The capacitors shown in Fig. 7.13 present a low impedance to interference currents and further reduce the proportion of energy passing through the filter. The optional earth-line choke, L_e, blocks high-frequency interference travelling along the earth line.

The performance of a filter can, in principle, be stated by giving the attenuation or insertion loss as a function of frequency. There is a difficulty, however, in that the performance of the filter depends on the impedances of the source and load networks and these vary from one application to another. Tests may be needed, therefore, to establish the performance of a particular filter in a given application.

Data and signal cables can be filtered at the point where they enter a system using filtered connectors with integral distributed filter elements. These can take the form of a capacitive sleeve connected to ground around each pin of the connector, or a lossy ceramic or ferrite tube with conductive coatings on the inner and outer cylindrical surfaces, behaving as a lossy transmission line.

For r.f. applications leadthrough or feedthrough capacitors are used to pass d.c. power through a bulkhead without allowing r.f. energy to pass.

At high frequencies the earth line cannot be considered as a negligible impedance to earth. High-frequency interference can therefore propagate along the earth conductor.

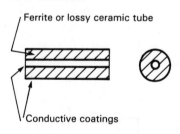

A distributed filter element.

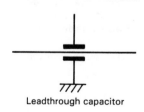

Leadthrough capacitor

Shielding

If we wish to exclude electromagnetic fields from a region inside an electronic system we can surround the region with an electromagnetic *shield*. Static electric fields are easily excluded from an equipment enclosure by a conducting screen or Faraday cage, but unfortunately, radiated electromagnetic fields are *time varying* and have both electric and magnetic components. A static electric field is always perpendicular to the surface of a conductor, cannot penetrate the conductor and cannot cause current to flow in the conductor. Time-varying electromagnetic

117

fields, however, *can* penetrate a conducting shield and *can* induce currents in the shield. The degree of penetration varies with the frequency of the field and the type of shielding material.

Shielding design is largely therefore a matter of understanding the mechanisms by which a shield works coupled with a knowledge of the type, frequency and magnitude of the incident field. Tabulated shielding effectiveness figures for various materials at a range of frequencies can then be consulted to select a material or combination of materials and to decide on the thickness required.

The shielding effectiveness, S, of a shield is defined as the ratio of incident power to power passing through the shield. S is made up from three components which add if expressed in decibels:

$$S = R + A + B \qquad (7.6)$$

R is a reflection loss due to an impedance mismatch between the shield and the incident field and depends on the shield material, the field frequency and the field impedance. A is an absorption loss and depends on the shield material and field frequency but not the field impedance. B is a correction term which accounts for reflection from the inner surface of the shield. R is significant for high-impedance and plane-wave fields so that a thin layer of, say, copper can provide effective shielding. R for copper at 10 MHz is about 100 dB for plane-wave fields. For low-impedance fields there is less of a mismatch between the shield and field impedances and S must be made up mainly from the absorption loss A.

Practical shields have weaknesses such as joints and openings which complicate estimates of shielding effectiveness. Slots in a shield, for example, can act as antennas and re-radiate energy from circulating currents inside the shield into the volume enclosed by the shield. Removable covers and lids create narrow slots where r.f. energy can pass through the shield, and may need to be sealed with *radio-frequency gaskets* of knitted wire mesh or conductive foam polymers.

Suppression

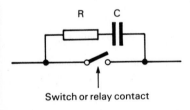

An R-C snubber network.

Motors and contact devices such as switches and relays can generate EMI in the form of transient or short-duration voltages, currents and electromagnetic fields, because of electrical arcing. Many types of switching circuits also generate transient interference. Empirical techniques exist to control or *suppress* this type of EMI including resistor and capacitor *snubber* networks across switch and relay contacts and suppression capacitor assemblies for suppression of motors.

Measures taken to reduce the high-frequency content of signals, such as limiting the rise-time of pulse edges, can also be regarded as suppression.

Applications Studies

This section presents studies of several applications where parasitic electrical and electromagnetic effects have an important influence on performance and design, both to illustrate the types of problems which can be met with and to show how good engineering practice can be justified in terms of the effects outlined so far in this chapter.

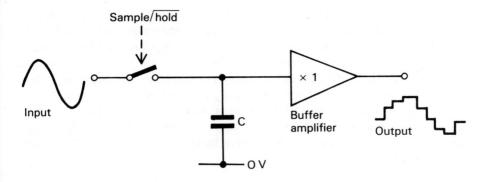

Fig. 7.14   A sample-and-hold (S/H) circuit.

A Sample-and-Hold Circuit

Figure 7.14 shows a common circuit in which parasitic effects are an important influence on performance. A sample and hold (S/H) circuit samples or captures the voltage at its input and holds that voltage at its output until the next sample is taken. They commonly precede analogue-to-digital converters (ADCs) in systems where an analogue signal, such as speech, is to be converted to digital form for long-distance transmission or signal processing. The operation of this circuit is, in principle, very simple. To sample the input signal, the switch, which is in practice a field-effect transistor (FET), is closed. The capacitor, C, charges up to the value of the input voltage and will *follow* the input voltage as long as the switch remains closed. To hold the sampled value, the switch is opened, isolating the capacitor C from the input and allowing the capacitor to hold the sampled value. A unity-gain buffer amplifier provides output current to supply a load, such as the input of an ADC, without drawing current from the capacitor. In an ideal S/H circuit the capacitor would charge instantaneously when the switch was closed and would hold the stored value precisely for an indefinite time when the switch was open. In a real circuit the time taken to charge the capacitor, or the *acquisition time*, is not zero and the charge stored on the capacitor leaks away when the switch is open. The output voltage is said to *droop* during the hold state. The droop rate, in mV s⁻¹, is an important performance parameter in high-resolution analogue-to-digital conversion applications. How can we estimate these parameters and what design steps could we take to improve the performance of this circuit?

Figure 7.15 shows the same circuit, but with some of the parasitic properties

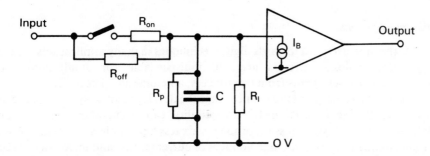

Fig. 7.15   The circuit of Fig. 7.14 with some of the parasitic circuit elements added.

added. The switch cannot be perfect and will have a small but non-zero resistance R_{on} when closed and a large but not infinite resistance R_{off} when open. The capacitor has a parasitic dielectric leakage resistance R_p which can be reduced by choosing a suitable type of capacitor, and the buffer amplifier has an input bias current I_B. All of these parasitic effects are properties of the components and can be reduced by careful choice of components. R_l, however, is a leakage resistance to 0 V due to the printed circuit board material, which can be increased by careful layout design to make leakage paths as long as possible and perhaps by using a solder resist to reduce moisture absorption by the PCB laminate, which would otherwise reduce R_l.

Worked Example 7.4

A precision S/H IC with 1 nF hold capacitor has a stated maximum droop rate of 30 mV s^{-1}. If the maximum input voltage is 5 V what is the total leakage current from the capacitor? The hold capacitor is connected to an external pin of the IC. What order of insulation resistance is needed on the PCB if the droop rate is not to be significantly increased?

We are assuming a constant leakage current, a valid approximation since the hold voltage is normally digitized by a following ADC within a few milliseconds.

Solution The capacitor voltage and the leakage current are related by the equation $i = C\, dv/dt$.

Since dv/dt is 0.03 V s^{-1} the leakage current must be 30 pA. At a maximum hold voltage of 5 V, this corresponds to a leakage resistance of 5 V/30 pA which is 167 GΩ (1.67×10^{11} Ω).

If the droop rate is not to be significantly increased by PCB leakage the insulation resistance between the hold capacitor and ground on the PCB must be, say, five times higher or about 800 GΩ.

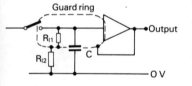

The leakage current through R_l can be reduced by adding a *guard ring* around the capacitor terminal as sketched in the margin. The guard ring is held at the same potential as the capacitor by connecting it to the buffer amplifier output. The leakage resistance R_l is split into two series resistances R_{l1} and R_{l2} as shown and since the potential difference across R_{l1} is zero to a close approximation, the leakage current that flows from the capacitor is negligible.

Exercise 7.4

If a guard ring is added to the S/H circuit in Fig. 7.14 by forming a ring from PCB track around the capacitor terminal, is it a good idea to use a solder-resist coating on the PCB or not?

Analogue Systems

Despite the modern trend towards digital techniques in many applications areas of electronics, analogue circuits are still indispensable for conditioning and amplifying low-level signals from transducers, antennae and detectors. New active devices and IC techniques tend to encourage the development of systems of greater sensitivity. Problems with EMI susceptibility can occur therefore in many analogue systems. EMC emission problems are less likely unless a system operates at radio-frequencies, or with high voltages or currents. R.f. and microwave circuit construction requires special techniques which have been excluded from this book.

Good engineering practice in analogue system design is to limit the operating bandwidth of the system, as already described. We should be aware, however, that measures taken to limit the bandwidth may not be effective at higher frequencies as the parasitic properties of components depend on frequency. High-frequency roll-off in an amplifier, for example, depends on strong negative feedback which may depend on a feedback capacitor. If the series inductance of the capacitor becomes significant at higher frequencies, the feedback provided by the capacitor will cease to hold down the amplifier gain.

Even when the bandwidth of a system has been limited to the minimum necessary, EMI falling outside the system bandwidth may be translated into an *in-band* signal by non-linearities in the system, in the same way that an amplitude-modulated r.f. carrier can be demodulated by a p–n junction. Two frequencies can also be combined by *intermodulation*, to produce sum and difference frequencies, by non-linear components acting as spurious analogue multipliers. In-band interference generated by frequency translation can be a serious problem.

Given that two voltage sources $V_1 \sin \omega_1 t$ and $V_2 \sin \omega_2 t$ are connected in series to a non-linear device with a characteristic given by $I = kV^2$, where k is a constant, show **Exercise 7.5**

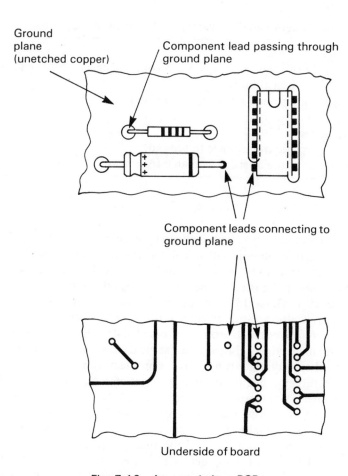

Ground plane (unetched copper)

Component lead passing through ground plane

Component leads connecting to ground plane

Underside of board

Fig. 7.16 A ground plane PCB.

that the device current will contain sinusoidal components at frequencies of $(\omega_1 + \omega_2)$ and $(\omega_1 - \omega_2)$.

Analogue circuits are often designed as *ground-plane* PCBs. Double-sided PCB construction is the simplest and cheapest method of building a ground-plane board. The component side of the board is covered all over with copper except where component leads pass through the board. This side of the board is the ground plane and serves as the 0 V reference for the whole circuit. Most, or all, of the tracks are laid out on the underside of the board. Figure 7.16 shows this type of construction and Table 7.1 lists the main advantages and disadvantages of a ground-plane board.

Careful attention to filtering, using R–C networks and ferrite beads on power-supply rails and bias networks and on signal paths between stages of a circuit, is essential to control power-line interference and to reduce the amplitude of EMI at the earliest possible stage in the system, before frequency translation and/or amplification occurs. In an amplifier system signal feedback from later stages to earlier stages must also be controlled, including feedback through power-supply rails by power-supply coupling as described earlier in this chapter.

Table 7.1   Advantages and disadvantages of ground-plane p.c.b.s

(a) Advantages
1.   Reduced parasitic capacitance between components and between tracks.
2.   Reduced EMI susceptibility.
3.   Low impedance ground return to power supply.

(b) Disadvantages
1.   Increased parasitic capacitance to ground.
2.   Only one side of the board is available for tracks.

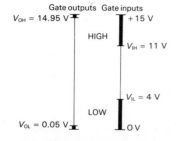

The sketch above shows the gate output and gate input logic levels for 4000 series CMOS logic. A worst-case logic low level at an output is 50 mV, which is 3.95 V less than the worst-case logic low threshold at a gate input. A spurious voltage of almost 4 V must therefore be added to a gate output in the low state to cause an error. This value is called a *noise margin*. A power-supply voltage of 15 V has been assumed.

Digital Systems

Digital electronic circuits are very widely used in computers, control systems, communications, instrumentation and signal processing. They often include microprocessors and other LSI circuits operating at clock frequencies of tens of megahertz and with pulse risetimes below 1 ns. A moderately sized PCB of double Eurocard format (233.4 × 160 mm) can accommodate a hundred or more ICs, sufficient to draw a power-supply current of several amperes. Digital PCBs and systems can therefore be troublesome sources of EMI, generating low-impedance, predominantly magnetic local fields and far-field emissions at frequencies of over 100 MHz. EMI susceptibility is not often a problem with digital systems because they are inherently immune to spurious voltages of several hundred millivolts or more. Conducted EMI entering a system is most likely to be dealt with adequately by measures such as filtering taken to control conducted EMI leaving the system.

Mixed analogue and digital systems such as microprocessor-based instruments can present a serious EMC problem if the analogue parts of the system are processing low-level signals as is frequently the case. Conducted and induced EMI from the digital circuits must be contained and controlled. Separate power supplies and grounds for the analogue and digital parts of the system are advisable, and care should be taken to segregate the analogue and digital circuits.

High-speed digital systems operating at clock frequencies of tens of megahertz and with pulse rise times below 1 ns are excellent examples of distributed-parameter systems where the finite speed of light is important. Consider for example, a single logic gate on a PCB. We have seen that the propagation velocity of electromagnetic signals on a PCB transmission line is about 150 mm ns^{-1}. Time delays in *signal* propagation are not likely to be a major problem unless our system is operating at a clock frequency of hundreds of megahertz, provided the system is reasonably compact. Time delays in *energy* propagation from the power supply to a logic gate can cause problems, however, if the gate switches rapidly compared to the propagation time from the power supply to the gate. Figure 7.17 illustrates the problem which can occur, especially with TTL logic. When the output of gate 1 is high a small reverse current of 40 μA flows into the emitter of the input transistor of gate 2. When the output of gate 1 switches to low, however, a current of 1.6 mA flows out of the input of gate 2 as shown. If gate 1 is driving the maximum allowable number of gates (10), gate 1 sinks 16 mA as shown. This current must eventually be supplied by the power supply, but initially the current must be supplied locally to gate 2 until a signal travels from gate 2 to the power supply *and back*. If there is no local source of current the rail voltage in the vicinity of gate 2 will fall, possibly causing spurious logic transitions in nearby gates. The problem is made worse because *both* output transistors of gate 1 conduct during the output transition, drawing a transient current from the supply rail. If a TTL logic circuit is to operate reliably, therefore, local reservoirs of charge are needed close to each logic gate in the form of *decoupling capacitors*. Traditionally, 100 nF ceramic capacitors are used, but there is no reason why other types should not be suitable provided charge can be extracted sufficiently rapidly from the capacitor. Within the last few years flat ceramic capacitors have been available for mounting underneath DIL IC packages, sharing the same PCB holes as the IC.

Supercomputers and high-speed signal-processing systems implemented in emitter-coupled logic (ECL) operate at clock frequencies of over 200 MHz, or a clock period of less than 5 ns. Differences in clock propagation times to separate parts of the system (clock skew) must be taken into account during design. The same problem can arise with a much slower digital system such as a data transmission network if the system is large (say distributed throughout a large building).

The term *decoupling* is historic and does not accurately describe the function of these capacitors. They are *charge reservoirs*.

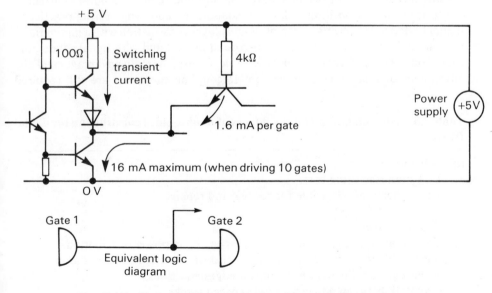

Fig. 7.17   The switching transient problem in logic circuits.

Table 7.2   Advantages and disadvantages of multilayer construction for digital PCBs

(a) Advantages
1. Low-impedance power distribution.
2. More uniform Z_0 for signal tracks.
3. Reduced EMI susceptibility because of low circuit loop area (all conductors are close to the ground plane).
4. Reduced EMI emission because of slower edges on logic transitions, caused by distributed capacitance, and low loop area.
5. Reduced crosstalk between tracks because of proximity to ground plane.

(b) Disadvantages
6. Higher manufacturing costs compared to double-sided PTH boards.
7. Difficult to rework and repair.

After local capacitors have supplied the transient current demand caused by a logic transition, the power supply must take over to supply the steady current and replenish the local reservoirs. The self-inductance of the power-supply loop now becomes important: a large open loop of power rail and return wiring will have a larger self-inductance than a small loop where supply and return conductors are both short and close together throughout their length, and the self-inductance of the large loop will restrict the increase in current and cause a voltage drop along the supply and return conductors. Depending on the distance to the power supply, there may be a need for a few larger reservoir capacitors on a PCB to supply current after the local decoupling capacitors have supplied the immediate transient demand but before the power supply has had time to respond.

The opposite situation to that described above occurs when there is a drop in current demand due to a logic transition: the reservoir capacitors have to absorb surplus current until the power supply responds and reduces its output current.

We are now in a position to look at examples of good engineering practice in digital PCB design. The first example is multilayer construction with a minimum of two internal layers for supply rail and ground planes, and with signal conductors routed on the outer faces of the board. Table 7.2 lists the main advantages and disadvantages of multilayer construction. Decoupling capacitors are still required

Table 7.3   Advantages and disadvantages of double-sided construction for digital PCBs

(a) Advantages
1. Low manufacturing cost.
2. All connections accessible for rework and repair.

(b) Disadvantages
3. Careful attention to layout needed, especially in power distribution.
4. Susceptible to radiated EMI because of open structure of circuit loops.
5. Significant EMI emissions from fast edges and clocks.
6. Crosstalk between adjacent closely spaced tracks.

(depending on the logic technology being used), even though the PCB itself contributes some distributed capacitance.

Estimate the total capacitance between power and ground planes of a double Eurocard (233.4 × 160 mm) multilayer PCB with a plane spacing of 0.3 mm and $\epsilon_r = 6$. $\epsilon_0 \cong$ 9 pF m^{-1}. Comment on the magnitude of the result. (*Answer*: about 7 nF)

Exercise 7.6

Multilayer PCBs are more expensive than double-sided boards and are also more difficult to rework and repair, so that double-sided PCBs are preferred for many applications. Figure 7.18 shows an example of good engineering practice in double-sided PCB design for a digital circuit and Table 7.3 lists the main advantages and disadvantages of double-sided construction. The power-supply distribution scheme illustrated is of interest. The supply and return conductors are run alongside each other and are interconnected in both directions to form a grid system. This reduces the inductance of the power-supply loop and also minimizes radiated interference as far as is possible without resorting to multilayer construction. Tracks on one side of the board run parallel to each other and the power-supply conductors and are perpendicular to those on the other side of the board. The tracks on the solder side run perpendicular to the direction of the IC packages to reduce solder bridging problems when the board is flow-soldered, so that the power-supply busbars running parallel to the ICs must be on the component side of the board, as illustrated. It remains to be seen how surface-mount boards using PLCCs and LCCs will be laid out.

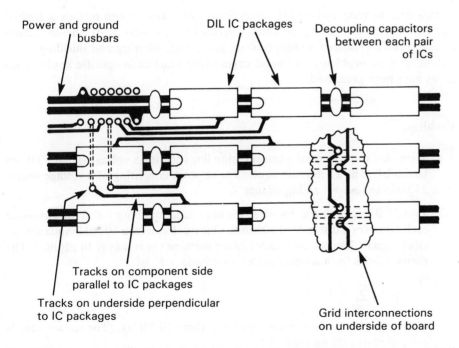

Fig. 7.18 An example of good practice in double-sided PCB layout for a digital system.

125

Summary

Many problems with electronic circuits and systems can be attributed to parasitic electrical and electromagnetic effects. The symptoms of these effects can include crosstalk, instability, pickup and interference. The causes of these symptoms include parasitic properties in components, parasitic circuit elements and electromagnetic coupling and radiation.

The finite speed of light, which is also the maximum speed at which electrical and electromagnetic signals and energy can propagate, influences the behaviour of an electronic system whose dimensions are greater than the speed of light multiplied by the signal period or time duration. Under these circumstances, lumped-parameter circuit models are no longer valid and conductors must be considered as transmission lines. The speed of signal propagation on a transmission line is $c/\sqrt{\epsilon_r}$, where c is the speed of light and ϵ_r is the relative permittivity of the transmission-line dielectric. There is no signal reflection from the end of a transmission line terminated with its characteristic impedance.

Unwanted electromagnetic energy entering a system is known as electromagnetic interference or EMI and can propagate by conduction along wires and cables or by radiation through free space. Parasitic antennae in electronic systems can radiate or receive electromagnetic energy. Close to an antenna in the near-field region the electromagnetic field may be predominantly electric or predominantly magnetic. Farther away in the far-field region the energy in the field becomes divided equally between the electric and magnetic components and the field impedance equals the impedance of free space. The study of electromagnetic interactions between electronic systems and sub-systems is known as electromagnetic compatibility or EMC.

Measures to reduce the EMI susceptibility of a system can include a limited bandwidth, well-designed grounding, filtering and shielding. EMI emissions from a system can be reduced by suppression, grounding, filtering and shielding.

Examples of problems and good engineering practice in specific applications areas have been presented.

Problems

7.1 Show that the output of a transmission line carrying a voltage signal $V_i(t)$ can be modelled as a Thévenin equivalent network consisting of a voltage source $2V_i(t)$ in series with an impedance Z_0.

7.2 Two PCB tracks with a characteristic impedance of about 100 Ω run alongside each other for a distance of 100 mm. One track carries a 10 MHz squarewave clock signal. The parasitic capacitance between the tracks is 10 pF m^{-1}. The Fourier series for a squarewave of amplitude $\pm V_p$ is

$$\frac{4}{\pi} V_p \sum_{n \text{ odd}} \frac{1}{n} \sin n\omega_0 t$$

where ω_0 is the fundamental frequency (here 10 MHz). (The squarewave is composed of odd harmonics.)

(a) What is the highest frequency at which the capacitive coupling between

[handwritten: a) $F = \frac{1}{t} = \frac{V}{S} = \frac{150\text{mm ns}^{-1}}{100\text{mm}}$]

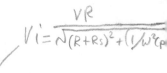

$$V\acute{i} = \dfrac{VR}{\sqrt{(R+R_S)^2 + (1/w^2 C_P}}$$

the tracks can be modelled as a lumped-parameter circuit? Assume a propagation velocity of 150 mm ns^{-1}.

(b) By treating the parasitic capacitance as if it were lumped, and the tracks as 50 Ω source and load networks, use eqn (7.1) to find the interference voltage, expressed in decibels relative to the squarewave amplitude, coupled from the clock track to the other track at all harmonics up to the 9th (90 MHz). Is this a fair approximation?

7.3 A data bus for a digital computer is designed so that each conductor has a characteristic impedance of 120 Ω. Open-collector logic is used to drive the bus and resistor networks are to be used at each end of the bus as shown to give a Thévenin source voltage of 3 V and a resistance of 120 Ω.

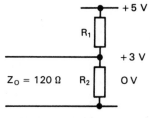

(a) What values of resistor are required?

(b) What is the maximum data rate achievable over a 3 m length of bus if the transmitter is at one end and the receiver is at the other end, and the receiver acknowledges each datum transferred? ($\epsilon_r = 6$).

7.4 An 8-bit bus driver IC drives eight 120 Ω transmission lines.

(a) If all eight gates drive their outputs to 3 V simultaneously what is the local instantaneous current demand?

(b) If the power supply is 300 mm away and signals propagate at 200 mm ns^{-1} what is the absolute minimum local capacitance required to prevent a fall in rail voltage of more than 50 mV?

(c) Why, in practice, is the answer to (b) optimistic?

8 Reliability and Maintainability

☐ To introduce reliability and maintainability.
☐ To discuss the meaning of the term *failure*.
☐ To introduce the 'bathtub' curve.
☐ To introduce quantitative measures of reliability and maintainability.
☐ To outline the principles of high-reliability systems.
☐ To discuss maintenance and design for maintainability.

The word *reliability* is used both in the sense defined here and also as the name of a subject or discipline concerned with failure and the analysis and prediction of failure.

Two useful general references for this chapter are Cluley and O'Connor. Many of the technical terms used in this chapter are defined in BS 4778.

Design for function was normal in the 1950s and early 1960s when electronic products were relatively expensive compared to the labour costs involved in repair. Today, electronic products are cheap compared to labour costs and the cost of repair is much more significant.

All electronic equipment has a limited *life*. The materials and components from which products are fabricated gradually deteriorate through wear or through physical and chemical processes such as creep, fatigue, diffusion, corrosion and embrittlement, leading eventually to *failure*. By replacing worn, aged and faulty components, equipment life can be extended, but eventually this becomes more expensive than total replacement.

It is possible to estimate the probability of a product continuing to function without failure over a specified time period or amount of use (the *reliability* of the product) by using data from *life tests* on samples of the product or from data on the reliability of the product's components. It is not, however, possible to predict the actual time of failure of an individual specimen since there is a degree of randomness inherent in failure mechanisms. Reliability is therefore a statistically based subject.

Modern electronic components, especially integrated circuits, are highly reliable, particularly when their complexity is taken into account. This has encouraged the widespread application of electronic systems in industry, offices and homes. In the past, many electronic products were designed to a functional or performance specification with little consideration of reliability or maintainability. This was possible because electronic systems of moderate complexity operating under benign conditions are inherently reliable. This, in turn, encouraged the development of more complex products and the introduction of electronics into more hostile environments where operating stresses are more severe. For example, road vehicles, where electronics have been applied to ignition and engine management systems, are a particularly arduous environment, providing vibration, extremes of temperature and electrical interference which can cause rapid failure of a poorly designed system. Electronics is also increasingly being applied in safety-critical applications such as life support, railway signalling and aircraft control systems (fly-by-wire). There is thus an increasing need for highly reliable electronic systems.

Improved product reliability and maintainability can be economically worthwhile. Reliable, easily maintained products have a lower *cost of ownership* partly because of the low maintenance costs but also because they are available for use for a greater proportion of the time than an unreliable product which requires extensive maintenance. The manufacturer of a product can also benefit from improved reliability and maintainability because he can reduce his support resources for the product by stocking fewer spare parts and by having fewer staff and less equipment devoted to repairs on faulty units. This is especially so if the product is large and complex and needs to be repaired by field service staff visiting the customer's premises or site. Complex products such as computer systems are increasingly being monitored remotely over telephone lines to ensure that scarce and expensive field service staff visit the customer's site equipped with the correct replacement units when a fault develops.

Failure

Failure of an electronic system or product normally means that the system is no longer able to perform one or more of its normal functions. Failure of an oscilloscope, for example, might mean that no trace was displayed for one of the vertical channels. A *fault* in a pocket calculator could result in loss of the divide function but not the add, subtract and multiply functions. *Complete failure* means that some function of the system is completely lacking, as in the example of the pocket calculator. Total lack of all function, perhaps caused by a fault in a system's power supply, would also be a complete failure, but a complete failure need not be total. A fault in a digital voltmeter could prevent measurement of voltage on one range but not on others. This would constitute a *partial failure* of the voltage measurement function. A more subtle failure is also possible in this case: suppose the voltmeter is accurate to $\pm 1\%$ when working correctly and a *gradual fault* occurs such that the voltmeter reading is in error by 5%. The voltmeter has failed even though it is still able to measure voltage. This type of failure is known as a *drift failure* and can occur in measuring systems and instruments wherever some performance parameter can drift over time. If the drift is consistently in one direction it is said to be *monotonic* and once the performance drifts outside an acceptable range the drift failure is permanent. If the drift is *non-monotonic* it is possible for performance to drift back into the acceptable range after a drift failure so that the failure is non-permanent.

Failure Mechanisms

Failures in electronic systems are usually attributable to failure of a component or electrical joint within the system. Component failure can be due to damage or stress during system assembly. A component might be overheated during soldering for example, or damaged by electrostatic discharge during handling. This type of failure is known as *infant mortality* and occurs soon after system assembly. Component failure can also be due to an inherent weakness or *flaw* caused by defective materials used in manufacture of the component or by abnormalities in the manufacturing process. This type of failure is known as a *freak failure* and can occur after thousands of hours of use. A third possible cause of component failure is *misuse*: a component may be subjected to stress beyond its normal ratings. This could be the consequence of a design error where the designer failed to consider behaviour of his system under all relevant conditions, or could be a *secondary failure* resulting from failure of another component in the system. A short-circuit fault in a power-supply reservoir capacitor, for example, could cause secondary damage to the bridge rectifier before the fault current was interrupted by a fuse.

In the calculator example, the failure of the divide function is a *complete failure* of that function, but not a total failure of the calculator. If a reciprocal key is available, continued use of the calculator is possible by using the reciprocal and multiply functions to implement division.

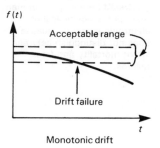

Monotonic drift

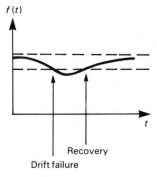

Non-monotonic drift.

A final failure mechanism is known as *wear-out*, and is due to component ageing or wear. Wear-out failures become more likely as a system becomes older. Ageing is due to physical and chemical processes which cause deterioration of a component over time, whereas wear occurs mainly in components with moving parts such as switches and potentiometers where mechanical abrasion or sparking gradually removes material from the contacting surfaces.

The 'Bathtub' Curve

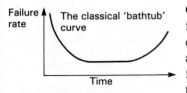

The classical bathtub curve illustrated here is given in many texts but is not typical of electronic systems because it does not distinguish between infant mortality and freak failures.

Consider a population of electronic products in mass production. If the number of failures per unit time, or *failure rate*, is plotted against time from manufacture, a curve resembling that shown in Fig. 8.1 is often found, and is known from its shape as the *bathtub* curve. There are three regions shown in the figure. In the early failure region there is initially a high failure rate due to infant mortality failures. As the number of infant mortalities falls, typically after 10 hours of operation, the failure rate rises again and then falls, due to freak failures which continue to occur at a declining rate up to 100 or 1000 hours of operation. The constant-failure-rate region normally lasts for five to ten years, during which a low incidence of freak failures occurs. This region is often referred to as the *useful life* or *service life* of the product. The total number of failures in these first two regions is normally a small fraction of the total population. Failures in these regions are normally repaired (assuming the product is accessible and designed for repair) and then continue to function throughout the constant-failure-rate region. The final part of the bathtub curve is known as the *wear-out* region. During this period, failure rate increases as components begin to fail through ageing and wear. Repairs may still be effected to keep a unit serviceable but eventually the cost of continued repair becomes uneconomic and the unit has to be withdrawn from service. All engineering products suffer this fate eventually.

The infant mortality failure rate can be reduced by *quality control* in manufacturing and assembly and by inspection and testing of components before assembly. Poor quality soldered joints, for example, might be avoided by regular

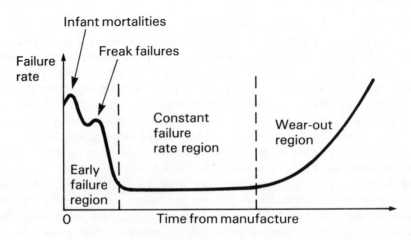

Fig. 8.1 The 'bathtub' curve.

inspection of the soldering machine or by testing sample boards run through the machine at intervals during a production run. The added cost of quality control must be balanced against the savings due to reduced incidence of failures and improved production yield.

If the products are *shipped* out of the factory directly after testing, the user or customer may experience early failures due to infant mortality. The added cost to the manufacturer of repairing or replacing the failed units can be avoided by *burn-in* testing: all units are tested under operating conditions in the factory for long enough to catch most of the early failures before they reach the customer. Mild stress such as heat, vibration and so on can increase the proportion of weak units detected in a given time, although there may be some reduction in the service life of the whole population as a result of the stresses. Detection of weak units in this way is known as *stress screening*.

The failure rate during the useful life of a product where the failure rate is fairly constant can be reduced by several means. One is to use high-reliability components which have been burned in by the component manufacturer so that early freak failures have been eliminated; another is to *derate* components, operating them well within their normal ratings. Resistors, for example, might be operated at no more than 50% of their normal power rating and at least 40°C below their normal maximum operating temperature. Diodes and transistors might be derated to 50% of their normal anode and collector currents respectively and all semiconductor devices might be limited to junction temperatures below 90°C (silicon). Further reductions in system failure rate can be achieved by redundancy and by minimizing the number of components for reasons which will be seen later in this chapter.

Wear-out failures can be postponed, but not prevented, by replacing worn or aged components before they fail. This is known as *preventive maintenance* and extends the useful life of a product. Some components may be replaced only if worn or shown to be aged by testing, but in many cases, components may be replaced after a fixed time or amount of use. On a moderate-sized electronic system which is not operated continuously, an elapsed time indicator or hour meter can be a useful and inexpensive way of recording actual operating time so that preventive maintenance can be carried out at the correct time.

Stress screening is an expensive process and is normally applied only to military and aerospace electronic systems. The subject is discussed further in Chapter 9.

Measures of Reliability and Maintainability

Reliability is a quantitative subject and there are several quantitative measures of reliability.

Maintainability is not easily quantified but the mean time taken to repair a faulty system can sometimes be a useful measure of maintainability.

Mean Time Between Failures (MTBF)

During the useful life of an electronic product the failure rate, usually denoted by λ, is approximately constant and is a useful measure of a system's reliability. The reciprocal of the failure rate is known as the *mean time between failures* (MTBF):

$$\text{MTBF} = \frac{1}{\lambda} \tag{8.1}$$

Eqn (8.1) is valid only if the failure rate, λ, is constant.

which is usually stated in hours. As an example of the level of reliability achievable with modern components, linear power supplies of up to 200 W rating can have MTBFs of more than 100 000 hours or 11 years. The corresponding failure rate is 10^{-5} hour^{-1}.

For non-repairable systems the *mean time to failure*, or MTTF, is quoted. In the case just given where the MTBF is 11 years, it is unlikely that the system will operate for another 11 years on average after a repair because of component ageing, and the figure given should properly be quoted as an MTTF.

MTTF and MTBF can also be defined in terms of the mean lifetime of a population of units and the mean interval between faults occurring in a population of repairable systems. If a sample of n units is tested or operated until all have failed, and the lifetime or time to failure τ is recorded for each unit, the MTTF is the mean lifetime.

The value given by eqn (8.2) is the *observed MTTF* for the population of n units. It can only be an estimate of the MTTF of the unit design, since the number tested can only ever be a sample of the total population.

$$\text{MTTF} = \frac{1}{n} \sum_{i=1}^{n} \tau_i \tag{8.2}$$

Exercise 8.1   A sample of 10 filament lamps is tested to failure with the following lifetimes in hours: 204, 1473, 650, 697, 1737, 558, 723, 215, 526, 1850. What is the observed MTTF? (*Answer*: 863 hours)

Mean Time to Repair (MTTR)

If a system is repairable, its *maintainability* can be quantified as the *mean time to repair*, or MTTR. Maintainability can be influenced by design, particularly in ease of access to components, by the method of repair, and the ease with which a fault can be diagnosed. Apart from design, MTTR can also be influenced by other factors such as geographical location of the system: the MTTR for a computer system located on a remote island will be longer than that for the same system located in a city near to the manufacturer's repair facility.

A system might be repaired by component replacement or by module replacement. Some computer systems now carry out self-diagnosis using diagnostic programs.

Availability

In some applications, high reliability is essential either because human life depends on continued operation or because repair is impossible or very expensive. In other applications, however, where repair is possible a more useful quantity for comparing one system with another is *availability*, defined as the proportion of time during which a system was, or is likely to be, functional.

Worked Example 8.1   Compare the availability figures for two computer systems. Machine 1 has a stated MTBF of 1000 hours and can be repaired by a service engineer in 3 hours on average. (His depot is over 1 hour's drive away.) Machine 2 has an MTBF of 500 hours and can be serviced by the manufacturers' on-site repair team in 45 minutes on average (by swapping modules).

Solution   The availability, A, can be calculated as

$$A = \frac{\text{MTBF}}{\text{MTBF} + \text{MTTR}}$$

Hence, for machine 1,

$$A = \frac{1000}{1000 + 3} = 99.70\%$$

for machine 2,

$$A = \frac{500}{500 + 0.75} = 99.85\%$$

Thus, although machine 2 breaks down more often, its availability is marginally (but not significantly) better.

Note: In a practical situation, the consequences of failure such as loss of data, and costs of repair would also have to be evaluated.

Reliability

MTBF is one quantitative measure of reliability, applicable where failure rates are constant. Another quantity, of more general applicability, is denoted by the word *reliability* itself. This is the *probability* that a device or system will function without failure over a specified time period or amount of use under specified operating conditions. For some devices and systems reliability is expressed per unit of time, whereas for other items reliability is stated per unit of use. A filament lamp, for example, might have an expected life of 1000 hours: the failure mechanism of the filament is dependent mainly on the time that the filament has been incandescent. A switch, on the other hand, could have a life stated as a number of operations because the failure mechanism is dependent on wear of the switch contacts.

In general, reliability, in the sense just defined, varies over time and is stated as a *reliability function*, $R(t)$, which gives the probability that an item will function without failure over time t. The value of $R(t)$ can be found, in principle, from life trials on a sample of items sufficiently large to give statistically valid results. The sample of items is tested to failure and the proportion of failed items is plotted as a function of time. The result is called the *lifetime distribution function*, $F(t)$. The slope of $F(t)$ is never negative since the proportion of failed items cannot decrease. $F(t)$ can also be interpreted as the proportion of items with lifetimes less than or equal to t. Then $1 - F(t)$ is the proportion of items with lifetimes exceeding t, which is also the probability that an item will function without failure over time t, or $R(t)$. Thus

It is assumed that all faults are permanent. This need not be so: faults which appear and then disappear are referred to as *intermittent*.

$$R(t) = 1 - F(t) \tag{8.3}$$

The slope of $R(t)$ is zero or negative for all values of t and for sufficiently large values of t, $R(t)$ becomes zero, because no item can continue to function indefinitely. These statements follow from eqn (8.3) and can also be justified from the definition of $R(t)$.

In practice life test data is often analysed by plotting $1/R(t)$ on a double logarithmic scale (ln ln $1/R(t)$) against lifetime on a logarithmic scale. This technique is known as Weibull analysis and is discussed by O'Connor.

Explain why the slope of $R(t)$ cannot be positive, using only the definition that $R(t)$ is the probability that a unit will function without failure over a time period t.

Exercise 8.2

Derivation of Failure Rate

The failure rate, λ, which in general varies with time, can be derived from the reliability function $R(t)$ introduced above, making no assumptions about the form of $R(t)$ and $\lambda(t)$ other than those stated above, namely that $R(t)$ must have zero or negative slope for all values of t.

Consider N items on test and represent the number of items failed as a function of time by $n(t)$. From the definition of the lifetime distribution function as the proportion of failed items

$$F(t) = \frac{n(t)}{N} \tag{8.4}$$

The rate of failure of each item is the rate at which items fail divided by the number of items still functioning at time t, which is $N - n(t)$.

$$\lambda(t) = \frac{1}{N - n(t)} \frac{dn(t)}{dt} \tag{8.5}$$

From eqns (8.3) and (8.4)

$$n(t) = N(1 - R(t)) \tag{8.6}$$

and differentiating with respect to t,

$$\frac{dn(t)}{dt} = -N \frac{dR(t)}{dt} \tag{8.7}$$

Substituting eqn (8.7) into eqn (8.5),

$$\lambda(t) = \frac{-N}{N - n(t)} \frac{dR(t)}{dt} \tag{8.8}$$

but from eqn (8.6)

$$\frac{-N}{N - n(t)} = -\frac{1}{R(t)} \tag{8.9}$$

and substituting from eqn (8.9) into eqn (8.8),

$$\lambda(t) = -\frac{1}{R(t)} \frac{dR(t)}{dt} \tag{8.10}$$

Equation (8.10) is a general relationship between failure rate and the reliability function $R(t)$.

Exercise 8.3 Show that an exponentially decreasing reliability function $R(t) = \exp(-kt)$, where k is a constant, corresponds to a constant failure rate λ equal to k.

A useful approximate relationship between $\lambda(t)$ and $F(t)$ can be derived from eqns (8.3) and (8.10) for small values of $F(t)$. Differentiating eqn (8.3) with respect to t,

$$\frac{dR(t)}{dt} = -\frac{dF(t)}{dt} \tag{8.11}$$

and rewriting eqn (8.10) in terms of $F(t)$,

$$\lambda(t) = \frac{1}{1 - F(t)} \frac{dF(t)}{dt} \tag{8.12}$$

Now, if $F(t) \ll 1$ (say $F(t) < 0.1$) then eqn (8.12) can be approximated by

$$\lambda(t) = \frac{dF(t)}{dt} \tag{8.13}$$

and the failure rate is approximately equal to the gradient of $F(t)$.

A batch of 5000 electronic ignition units was monitored during their first year in service. The number of failures per month was as follows:

Worked Example 8.2

Month:	1	2	3	4	5	6	7	8	9	10	11	12
Failures:	0	1	0	9	9	3	8	28	15	3	7	2

Plot the lifetime distribution function and use eqn (8.13) to produce a plot of failure rate.

Solution The cumulative number of failures each month and the proportion of failed units, $F(t)$ are:

Month:	1	2	3	4	5	6	7	8	9	10	11	12
Cumulative failures:	0	1	1	10	19	22	30	58	73	76	83	85
$F(t)$ (%)	0	0.02	0.02	0.20	0.38	0.44	0.60	1.16	1.46	1.52	1.66	1.70

Time in this example, as in any practical case, is not continuous but is divided into *discrete* periods (months in this case). Strictly $F(t)$ should be replaced by $F(n)$ where n is month number.

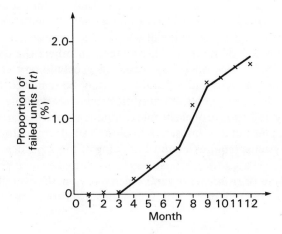

Fig. 8.2   Plot of $F(t)$ for the data in Worked Example 8.2.

The proportion of failed units, $F(t)$, is plotted in Fig. 8.2. Approximate slopes have been drawn and from these the failure rate $\lambda(t)$ has been plotted in Fig. 8.3 using the approximation of eqn (8.13). These are all early life (freak) failures — a car is not used continuously and the normal service life region of the bathtub curve is not reached until the second year of use.

Typically a car averages 12 000 miles per year. Assuming an average speed of 40 m.p.h. this represents about 300 hours running. The distinction between electronic systems which run continuously and those that are used intermittently is an important one from a reliability viewpoint.

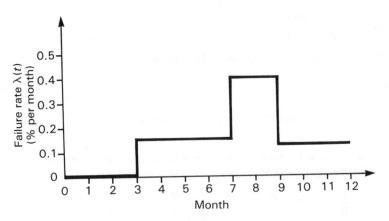

Fig. 8.3   Plot of failure rate $\lambda(t)$ for the data in Worked Example 8.2.

Exercise 8.4   Plot the failure rate data for Worked Example 8.2 directly from the number of failures per month and compare the result with Fig. 8.3.

High-reliability Systems

In some applications of electronic systems high levels of reliability, or low probabilities of failure, are required. Some examples are control systems in petro-chemical plants and nuclear power stations, railway signalling systems, instruments and control systems on board aircraft, and satellite communication systems. In some of these examples high reliability is required because human life depends on correct functioning of the electronic system, in others the consequences of failure are financial losses on a huge scale: a geostationary communications satellite cannot be salvaged and a failure can only be remedied by launching a replacement satellite. Many applications of electronics are possible only because of the inherently low failure rate of electronic components, particularly ICs which are functionally complex but are almost as reliable as a single transistor. The design of high-reliability systems requires an understanding of the dependence of a system's reliability on the reliabilities of its sub-systems and their components. This is achieved by analysis of *reliability models* which represent the overall reliability of a system in terms of the reliability of its components and sub-systems.

Satellites in low Earth orbit have been recovered by the US Space Shuttle. The cost of the failure is still enormous. An excellent example of a successful high-reliability system is the pair of Voyager space probes which took over ten years to travel to the outer planets of the Solar System. A brief article on the reliability aspects of Voyager appeared in *IEEE Spectrum*, 68–70, **18**(10), October 1981.

Series Systems

The simplest reliability models are of systems where every component must be functional for the system to be functional: failure of any component implies failure of the system. Many electronic systems with no special provisions for high relia-bility are of this type: every component has a functional purpose and all are essential if the system is to work correctly. Figure 8.4 is a diagram of the reliability model for such a system. For reliability purposes, each component is in series with all the other components. As a simple non-electronic example of such a system, consider a suspension bridge: the two main cables from which the bridge deck is

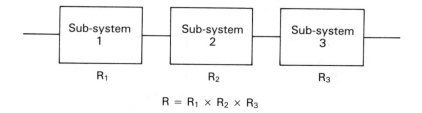

$$R = R_1 \times R_2 \times R_3$$

Fig. 8.4 A series reliability model.

suspended are a series reliability system, even though they are functionally in parallel, since failure of either cable would cause collapse of the bridge.

The overall reliability of a series system is the product of the reliabilities of the sub-systems. For the system shown in Fig. 8.4, if the reliability of sub-system 1 is R_1, 2 is R_2, 3 is R_3, then the reliability of the system as a whole is $R_1 \times R_2 \times R_3$. In general, when a series system has many more than three sub-systems with differing individual reliabilities, the overall reliability of the system is dominated by the reliability of the least reliable sub-system. The reliability of a series system is generally better if the system has as few components as are necessary for the purpose.

R_1, R_2 and R_3 are probabilities that the sub-systems will continue to function over a specified time period or amount of use. The overall reliability is the *compound probability* that all three sub-systems will continue to function. Probability is discussed by Cluley.

Show that the reliability of a series system cannot exceed the reliability of the least reliable component.

Exercise 8.5

Parallel Systems

In contrast to the series reliability system just discussed, in a parallel system such as that shown in Fig. 8.5, not all of the parallel units need be functional for the system as a whole to be functional. Many cars are now equipped with dual-circuit braking

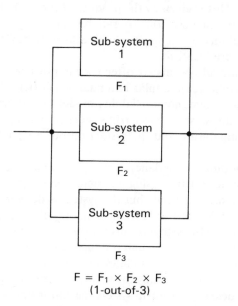

$$F = F_1 \times F_2 \times F_3$$
$$(1\text{-out-of-}3)$$

Fig. 8.5 A parallel reliability model.

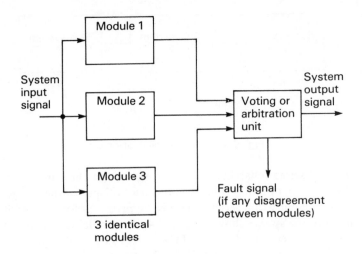

Fig. 8.6 A triplicated 2-out-of-3 system.

systems and can be brought to a halt even if one of the brake circuits has been damaged. Commercial airliners are able to fly with only two out of four engines working and the control systems for ailerons, flaps and rudder are at least duplicated and possibly triplicated to achieve the high level of reliability necessary for safety.

In a simple parallel system any one of the sub-systems must be functioning for the system as a whole to be functioning. This is called a *1-out-of-n* system. If we use the probability of failure, F, of each sub-system, where $F = 1 - R$, then the probability of failure of the total system is the probability that all n parallel sub-systems have failed. This is given by the product of the probabilities of failure of the sub-systems. Qualitatively, the overall reliability of a *1-out-of-n* system depends on the reliability of the most reliable sub-system. Often, of course, the parallel sub-systems are identical.

More elaborate parallel systems are often used in practice and Fig. 8.6 shows a scheme known as triplicated modular redundancy (TMR). (This is a functional block diagram of the system, not a reliability model.) A system of this type might be used in a nuclear power station as a trip system for a reactor: the system input could be reactor temperature and the system output a signal to shut down the reactor.

In a triplicated modular redundancy system, three identical electronic units or modules perform some function using a system input signal. All three modules normally produce identical outputs but the system is designed to work correctly even if one of the modules fails and produces an output differing from that of the other two modules. A TMR system is thus a *2-out-of-3* parallel system. A simple, highly reliable, voting or arbitration unit produces a system output based on agreement between any two of the three modules and also generates a fault signal if the three modules do not agree. Normally the voting unit is designed to be *fail safe*. In the example of the nuclear reactor trip system the voting unit would be designed to signal a reactor shut-down if no two modules agreed or if the voting unit itself was faulty.

Derive the reliability, R, of the 2-out-of-3 TMR system shown in Fig. 8.6.

Solution Let the reliability of the voting unit be R_v and the reliability of the triplicated modules be R_m, and the corresponding probability of failure of a triplicated module be F_m, where $F_m = 1 - R_m$.

The system can be modelled as a series system of two elements as sketched in the margin (this is a reliability model). The overall reliability of the TMR system using the rule for a series system is

$$R = R_{tm} \times R_v$$

where R_{tm} is the reliability of the triplicated module sub-system. R_{tm} is the probability that at least two of the redundant modules are functioning. $F_{tm} = 1 - R_{tm}$ is therefore the probability that two or all three of the redundant modules fail. There are three ways in which two out of the three modules can fail and one way in which all three can fail:

Probability that modules 1 and 2 fail and 3 does not fail	$= F_m^2 R_m$
Probability that modules 2 and 3 fail and 1 does not fail	$= F_m^2 R_m$
Probability that modules 1 and 3 fail and 2 does not fail	$= F_m^2 R_m$
Probability that all three modules fail	$= F_m^3$

Hence $F_{tm} = \quad\quad\quad\quad\quad F_m^3 + 3F_m^2 R_m$

Since $R_{tm} = 1 - F_{tm}$ the overall reliability of the TMR system is

$$(1 - F_m^3 - 3F_m^2 R_m)R_v$$

which can be shown to be

$$(3R_m^2 - 2R_m^3)R_v$$

Show that $1 - F_m^3 - 3F_m^2 R_m = 3R_m^2 - 2R_m^3$

Compare the reliability of a TMR measurement system with a non-redundant system using the same measuring module design if the module reliability is 0.999 and the voting unit is ten times more reliable than the measuring module.

Solution The voting unit has a reliability R_v of 0.9999 because its probability of failure F_v must be ten times less than the module probability of failure F_m, which is 0.001. Therefore the reliability of the TMR system from Worked Example 8.3 is

$$(3(0.999)^2 - 2(0.999)^3)0.9999 = 0.99990$$

and the probability of failure is 10^{-4}. This is a factor of ten better than the reliability of a single module.

Diversity

A possible weakness in parallel, redundant systems is the so-called *common-mode failure*, where all redundant units fail simultaneously from the same cause. As an example, if all the engines on a multi-engined airliner were supplied with fuel from the same fuel pump, a failure of the fuel pump would stop all the engines. However, even if separate fuel pumps are provided, there is still the possibility of a common-mode failure under some unlikely operating conditions because all the pumps are of the same design and all could fail at the same time if the unlikely condition occurs. This problem is especially severe with engineering software since identical copies of a program will always give identical results no matter what the input data. In highly critical systems, therefore, there may be a need to employ redundant parallel systems of different design to each other. Such an approach can be very expensive, especially in software-based systems, where the cost of developing one system can be high enough: a *diverse redundant system* requires two design teams developing software for two different computers and maybe even programming in two different languages.

An alternative for small software systems is to prove the software to be correct mathematically. Much research effort has been expended on the problem of proving the correctness of both programs and complex logic systems such as microprocessors. These techniques have so far been applied only in limited cases.

Maintenance

Maintenance is defined in BS4778 as

> The combinations of all technical and corresponding administrative actions intended to retain an item in, or restore an item to, a state in which it can perform its required function.

Some electronic products are designed for *nil maintenance* because of inaccessibility once in service or because it is cheaper to scrap a failed unit than to repair it.

Sometimes the word *servicing* is used to refer to these activities, and in many organizations the department responsible for carrying out maintenance is called the *servicing department*. Maintenance activities include *calibration* (adjusting a system so that its characteristics are within a specified range); *fault diagnosis* (establishing the cause of a failure); *repair* (removing and replacing faulty, damaged or worn parts and components) and *testing* (checking that a system is functioning correctly, especially after repair). The design of an electronic system can have a significant effect on the cost of performing these activities and often quite simple changes at the design stage can greatly reduce the cost of maintenance.

Maintenance Strategies

There are many different approaches to the maintenance of an electronic product dependent on the nature and value of the product, the type of manufacturer, the type of user and the desired level of reliability. The first decision to be made about a product is whether it is to be maintained at all. Some products are inherently non-maintainable, others are not economically maintainable. As examples, an IC cannot be repaired and it is unlikely that a cheap digital wrist-watch can be repaired for less than the original cost. Secondly, if the product is to be maintained, there is a choice between maintenance *in the field* (at the user's premises) or by returning the product to a service centre. In some cases the choice will be determined by the nature of the product and the type of maintenance needed. Large composite systems may be maintained by a combination of the two approaches, with major sub-assemblies maintained in the field and smaller units returned to a service centre for maintenance. Finally, there is a choice between *preventive maintenance* or the replacement of items before they wear out and *corrective maintenance* as needed to repair faults. Preventive maintenance can be applied only where components are liable to fail through wear-out. The majority of electronic components, especially semiconductors, have a useful life far in excess of the likely life of a system, so that failures due to these components are rare, per component, and are randomly distributed in time. Preventive maintenance cannot therefore reduce the failure rate attributable to these components.

Fault Diagnosis

An important and often difficult part of the process of repairing an electronic system after a failure is that of diagnosing the fault, or establishing what repair action is necessary to restore the system to its normal functioning state. Some faults, by their nature, are obvious and require little diagnosis. A failed indicator lamp, for example, can be seen to have failed and can be replaced. Most faults in electronic equipment, however, are much more subtle, often manifesting themselves by obscure symptoms. These types of fault often require considerable skill to diagnose and may require the use of expensive and sophisticated test equipment, such as logic analysers. Computers, being among the most complex of electronic systems, often present difficult problems in fault diagnosis.

If an electronic system is sufficiently complex to have been designed in modular form, the first step in diagnosing a fault is to establish which module is faulty. A minicomputer, for example, might be constructed from separate PCBs or boards including a processor board, memory boards and input–output boards. The symptom of the fault may be that the computer outputs garbled characters to a printer. As a first step in diagnosing this fault, the input–output board connected to the printer could be replaced by a spare identical board. If this cured the fault, the fault is most likely to be on the replaced board, otherwise some other part of the system might be suspected and replaced. This stage of fault diagnosis by modular replacement can be carried out by someone with little knowledge of the detailed workings of each module. One problem, however, is that spare modules are needed for substitution, unless the system contains more than one of each module type.

It is not certain that the fault is on the replaced board since the actual fault could be in the board connector, in which case removing and re-inserting the original input–output board might cure the fault.

Diagnosis of faults on PCBs to component level is more difficult than isolating a fault to a single module, since a detailed understanding of the circuit is necessary, together with skill and experience in fault-finding. The designer of a board can make fault-finding easier by adding *test-points* or special terminals to allow easy connection of test equipment. Also, if components are mounted in sockets, faults can be diagnosed by swapping suspect components. Many other possibilities exist for aiding testability and fault diagnosis by design.

See Wilkins for examples in the field of logic design.

Design for Maintainability

Attention must be given to the maintainability of a product at the design stage if maintenance costs are to be kept low. Clearly, if a product is to be designed for nil maintenance, construction techniques can be used which would not be acceptable on a maintained product. IC sockets, for example, are not worth using if the PCB on which they are mounted is to be a throw-away, non-repaired item. Assuming that we are considering only maintainable products here, what steps can be taken in design to make maintenance easier?

Perhaps the most important aspect of product design for maintainability is *access*: how easy is it to get at those parts of the product which require replacement, adjustment or inspection? Covers and panels should be easily removable and preferably free of attached wiring. Printed circuit boards should have connectors rather than permanently soldered wiring to simplify removal and replacement. Wiring to switches and so on can be fabricated with crimped push-on connectors, both to simplify manufacture and to make switch replacement easier.

Adjustable components such as preset potentiometers and trimmer capacitors should be used as little as possible, because of the time taken to set them during manufacture and because of the need to adjust them during maintenance. Where they cannot be avoided because the required circuit properties cannot be achieved with fixed components, they should be readily accessible and capable of adjustment while the equipment is operating. A test point or output should be provided so that the effect of the adjustment can be monitored.

Summary

Electronic systems have a limited life due to deterioration of components and materials with time. Components can fail due to inherent weakness, misuse and wear-out and can cause partial or complete failure of the system of which they are a part. The bathtub curve represents a typical pattern of failure rates for electronic systems characterized by a relatively high early failure rate as weak components fail (infant mortality and freak failures), a low fairly constant failure rate thereafter, and eventually an increasing failure rate due to component wear-out. The bathtub curve explains why electronic products are often tested for a period after manufacture (burn-in) and how the operating life of a system can be extended by replacing components before they fail due to wear-out.

Mean time between failures (MTBF) is a common measure of reliability and is the reciprocal of failure rate, λ, provided λ is constant. A general reliability function $R(t)$ can be defined even if failure rate varies with time. $R(t)$ is the probability that an item will be functioning at time t. $R(t)$ can be measured by determining the lifetime distribution function $F(t)$ from a life test on a sample of items. $R(t)$ is $1 - F(t)$. The failure rate $\lambda(t)$ can be expressed in terms of $R(t)$.

Where high reliability is needed, parallel redundancy can be applied to overcome the limited reliability of systems where every component must function for the system as a whole to function. Reliability models can be used to predict the reliability of redundant and non-redundant systems using simple probability methods. To overcome the problem of common-mode failure, diverse redundant systems of different design can be used.

Maintenance activities include calibration, fault diagnosis, repair and testing. Attention to maintainability at the design stage is essential if unnecessary maintenance costs are to be avoided.

Problems

8.1 Over a 6 month period a computer installation suffered four failures taking respectively 1, 2, 1.5 and 2 hours to repair. Also the system was shut down on 2 occasions for routine preventive maintenance for 2 hours each time. What are (a) the availability, and (b) the MTBF over this period?

8.2 A designer requires a power supply with a reliability of 0.8 over 2 years. (a) Assuming constant failure rate what MTBF is required? (b) The best suitable power supply available has an MTBF of 50 000 hours. If two of these are used in parallel with a number of diodes to isolate a failed unit what reliability must the diodes have?

See Exercise 8.3.

8.3 An instrument system for a chemical plant has two redundant measuring channels with an MTBF of 100 000 hours sharing a power supply with an MTBF of 150 000 hours. What is the probability of a system failure causing loss of the measurement function within 5 years?

8.4 An electronic unit has an MTBF of 100 000 hours and operates continuously. Assuming a constant failure rate

$$(R(t) = \exp(-\lambda t))$$

calculate the age at which this unit should be replaced if its reliability is not to fall below 75%.

9 Environmental Factors and Testing

Objectives
☐ To discuss a selection of common environmental factors which influence the performance and reliability of electronic systems.
☐ To discuss type testing and electronic production testing.
☐ To introduce design for testability.

Environmental Factors

This book opened with the statement that modern electronic products are to be found in a wide range of applications environments. These products must not only be fit for their intended purpose; they must also be able to survive the *stresses* imposed by their operating environments during their design lives. Some environments are relatively benign and impose little stress on an electronic system. A domestic living room, for example, provides a controlled-temperature, low-humidity environment reasonably free from mechanical stress and electromagnetic fields. Other environments, such as a road vehicle engine compartment, are much more hostile and subject an electronic system to wide variations in temperature, high humidity, mechanical shock, vibration and grime. The effects of many environmental factors are not easy to predict and indeed the presence of the factors themselves may not be foreseen at the design stage. There is thus a need for *testing* in simulated environmental conditions and in some cases full *field trials* in the actual operating environment to establish that the product will work as intended.

We can now discuss a selection of common environmental factors.

Temperature

The temperature of the operating environment is an obvious factor to be considered during design. There are three likely temperature ranges over which an electronic product might be required to operate, listed in Table 9.1. These are broad generalizations and in most applications some variation of these ranges might be encountered.

Circuit design for a wide operating temperature range is made difficult by the

Table 9.1 Operating temperature ranges for electronic equipment

Designation	Range (°C)	Environments
Commercial	+5 to +40	Homes, offices and laboratories
Industrial	0 to +70	Process plants, factory floors
Military	−55 to +125	Weapons systems, fighting vehicles

temperature dependence of many component parameters. Examples include the common emitter current gain, β or h_{fe}, of bipolar transistors, the input offset voltages of operational amplifiers, the breakdown voltages of Zener diodes, and of course, the values of resistors and capacitors, which were discussed in Chapter 5.

Worked Example 9.1

A weighing machine is required to weigh 0–25 kg in 10 g steps and to give a numerical readout. What is the precision needed in the machine's analogue-to-digital converter? The designer decides to use a 5.1 V BZY88 Zener diode to provide the voltage reference for the ADC. The diode has a temperature coefficient of breakdown voltage of 1 mV $°C^{-1}$. If the operating temperature range is $+5°C$ to $+40°C$ what would be the contribution from the Zener voltage reference to the worst-case error in the machine's reading if it was calibrated at 20°C?

Solution   10 g in 25 kg is 1 part in 2500 or 0.04%. (An ADC of 12-bit precision would be needed.) The worst-case temperature is 40°C or a change of 20°C from the calibration temperature. The Zener diode voltage will change by 20 mV from its 20°C voltage or a change of 0.4%. The error caused by the temperature variation of the Zener diode voltage is thus ten times greater than the precision required (and all other factors contributing to the error have been ignored). Hence either the specification has to be relaxed or an improved voltage reference used.

Temperature variation of component parameters can be a significant problem even for systems of modest performance operating over the commercial temperature range. Many circuit techniques have been devised to reduce the effect of temperature by adding compensating components. In the case of crystal oscillators, which are often used as precision frequency references, the resonant frequency of the crystal varies with temperature. To overcome this problem the oscillator may be housed in a small temperature-controlled enclosure or 'oven' in order to reduce the variation of output frequency with ambient temperature.

Clearly, all components within an electronic system must have an operating temperature range covering at least the operating temperature range of the product itself, unless some controlled-temperature environment is provided within the system.

Particular problems exist with the selection of components for operation at sub-zero temperatures. At these temperatures common difficulties include some battery types, for example, lead acid, which do not work well below 0°C, and TTL logic ICs which are rated down to only 0°C.

Battery behaviour at low temperatures was discussed in Chapter 4. Early liquid crystal displays (LCDs) had a slow response and exhibited poor contrast below 5°C. Improved materials have now reduced the minimum operating temperature for LCDs to around −30°C.

Exercise 9.1

TTL 74 series logic has an operating temperature range of 0–70°C, yet CMOS 4000 series logic can operate down to −40°C. Why?

Temperature and Reliability

A less obvious, but very important, effect of temperature is a reduction in the reliability of a product when operating at an elevated temperature or when subjected to repeated variations in temperature (thermal cycling). One way in which temperature and temperature cycling reduce reliability is the creation of

Possible solutions to this problem are: mounting the ceramic chip carriers in sockets; using a special-purpose PCB laminate with a TCE matched to that of the ceramic chip carrier; mounting the chip carriers on a ceramic substrate.

mechanical stresses within components and joints by differential thermal expansion, which may lead to fracture. This problem occurs with ceramic chip carriers soldered to epoxy–fibreglass PCBs, for example, where the temperature coefficient of expansion (TCE) of the ceramic (about 6 p.p.m. $°C^{-1}$) and the epoxy–fibreglass (about 12–16 p.p.m. $°C^{-1}$) are sufficiently different to make this method of mounting unreliable because of solder joint fracture after temperature cycling.

Worked Example 9.2

The largest chip carriers have 124 pads, 31 per side, spaced at 1.27 mm centres. The distance between pad centres at the ends of a side is 30 × 1.27 or about 38 mm. Over a 50°C temperature range the relative change in end-pad spacing between the chip carrier and the board is 38 mm × 50°C × 6–10 p.p.m. $°C^{-1}$ or 11–19 μm. This is about 1–1.5% of the pad spacing of 1.27 mm.

Similar problems with differential thermal expansion occur in plastic IC packaging, both between the silicon chip and the metal leadframe to which the chip is bonded and between the metal leadframe and the moulded plastic package. The TCEs of the leadframe and the plastic have to be matched to the TCE of silicon, which is about 2.6 p.p.m. $°C^{-1}$, to prevent fracture of the bond between the chip and the leadframe and to reduce moisture penetration along the leads.

A second, very significant effect of temperature on reliability is an increase in the rate of component ageing or deterioration with increased temperature. Many component wear-out failures are caused by chemical reactions occurring inside the component, and because the rate of reaction increases with temperature the life of a component is reduced by operating at higher temperatures. Many chemical reactions obey the Arrhenius equation

The Arrhenius equation is discussed by O'Connor, and in its chemical context in any 'A' level physical chemistry text.

$$\lambda = K \exp(-E/kT) \tag{9.1}$$

where λ is a reaction rate or failure rate, K is a constant for any particular reaction or component type, E is an activation energy for the reaction, k is Boltzmann's constant (1.38×10^{-23} J K^{-1}) and T is absolute temperature. The activation energy for a chemical reaction is the minimum energy required by the reactant atoms or molecules for the reaction to occur. Many reactions have activation energies between 0.5 and 1.5 eV or 0.8 and 2.4×10^{-19} J.

An electron-volt (eV) is the energy imparted to a unit charge (the charge on an electron) when transferred through a potential difference of 1 V. Since the electron charge is about 1.6×10^{-19} C, 1eV is equivalent to 1.6×10^{-19} J.

Worked Example 9.3

Calculate the increase in failure rate at a temperature of 60°C (333 K) over the failure rate at 20°C (293 K), assuming the Arrhenius equation (eqn (9.1)) applies and the activation energy for the underlying chemical process is 0.8 eV, or about 1.3×10^{-19} J.

Solution Given two temperatures T_1 and T_2 ($T_2 > T_1$) eqn (9.1) can be applied at both temperatures:

$$\lambda_2 = K \exp(-E/kT_2), \qquad \lambda_1 = K \exp(-E/kT_1)$$

Dividing one equation by the other eliminates K:

$$\frac{\lambda_2}{\lambda_1} = \frac{\exp\left(-E/kT_2\right)}{\exp\left(-E/kT_1\right)} = \exp\left(\frac{E}{k}\left(\frac{1}{T_1} - \frac{1}{T_2}\right)\right) \qquad (9.2)$$

and substituting the values of E, k, T_1 and T_2 gives

$$\frac{\lambda_2}{\lambda_1} = \exp\left(\frac{1.3 \times 10^{-19}}{1.38 \times 10^{-23}}\left(\frac{1}{293} - \frac{1}{333}\right)\right) = 48$$

The failure rate therefore increases by a factor of about 50 for a temperature rise of 40°C at this activation energy.

The increase in failure rate at higher temperatures has two important consequences. Firstly, long-term reliability tests can be speeded up by operating the devices under test at elevated temperature, yielding results in a shorter time of perhaps weeks or months rather than years, and therefore reducing the costs of testing considerably. This is called *accelerated life testing*. Secondly, product reliability is improved if components are kept cool. Power-dissipating components such as resistors and semiconductors can be derated by using a component with a higher power rating than is strictly necessary for the purpose. This derating will result in the operating temperature of the component being reduced with a consequent increase in reliability. Power semiconductors can also be operated at lower temperatures by attaching them to more effective heatsinks, or by providing forced cooling. In applications where electronic sub-systems form part of a larger system, operating temperatures may be reduced by careful siting of the electronics away from sources of heat.

A capacitor manufacturer quotes lifetimes for a range of capacitors as 120 000 hours at 40°C and 6000 hours at 85°C. If the failure rate is assumed constant and therefore proportional to the reciprocal of the component lifetime what is the activation energy, E, and what would the lifetime be at 100°C by extrapolation?
(*Answer*: 0.64 eV, 130 hours)

Exercise 9.2

A single activation energy is assumed here for the underlying chemical process.

Water and Humidity

Water is an obvious environmental hazard to any electronic equipment intended to operate out of doors or in a wet industrial environment. As all water except the very purest is conductive, short circuits, leakage currents and high-voltage flashover will occur if water comes into contact with electrical or electronic circuits. There is thus a need to design equipment enclosures to prevent the ingress of water. BS5490 defines 9 categories of protection against water ranging from 0 (no protection) to 8 (protected against submersion), and including varying levels of dripping and spraying water in between. There are standardized empirical tests defined for each category. Ideally, of course, every equipment enclosure for use in anything except a dry environment would be watertight, but totally sealed enclosures are expensive and may create problems with heat management and accessibility for maintenance. In practice, therefore, an equipment enclosure will be designed with the minimum degree of protection against water necessary for the application environment.

Humidity, or water vapour, is a less well-defined problem for the equipment designer. Humidity can be absorbed by some electronic engineering materials such

Engineering designers are often faced with the need to balance the benefits of a design factor against its cost and its technical disadvantages. This is known as a *trade-off*. Quite often, the factors to be traded off are not easily quantifiable, as here in the case of accessibility, and the engineer must exercise judgment based on experience.

Table 9.2   Degree of protection indicated by the second characteristic numeral of the IP classification: protection against water.

Second characteristic numeral	Degree of protection	
	Short description	Definition
0	Non-protected	No special protection
1	Protected against dripping water	Dripping water (vertically falling drops) shall have no harmful effect
2	Protected against dripping water when tilted up to 15°	Vertically dripping water shall have no harmful effect when the enclosure is tilted at any angle up to 15° from its normal position
3	Protected against spraying water	Water falling as a spray at an angle up to 60° from the vertical shall have no harmful effect
4	Protected against splashing water	Water splashed against the enclosure from any direction shall have no harmful effect
5	Protected against water jets	Water projected by a nozzle against the enclosure from any direction shall have no harmful effect
6	Protected against heavy seas	Water from heavy seas or water projected in powerful jets shall not enter the enclosure in harmful quantities
7	Protected against the effects of immersion	Ingress of water in a harmful quantity shall not be possible when the enclosure is immersed in water under defined conditions of pressure and time
8	Protected against submersion	The equipment is suitable for continuous submersion in water under conditions which shall be specified by the manufacturer. *Note:* Normally, this will mean that the equipment is hermetically sealed. However with certain types of equipment it can mean that water can enter but only in such a manner that it produces no harmful effects

Extracted from BS 5490:1977 with the permission of the British Standards Institution

as PCB laminates, causing reduced insulation resistance and increased leakage currents, which can be a problem in high-impedance circuits such as charge-sensitive amplifiers and sample-and-hold circuits.

The quantity of water vapour present in the air is given by the *relative humidity* (RH) which is the ratio of the density of water vapour in the air to the density of saturated water vapour in air at the same temperature. RH can be up to 100%, at which point water condenses out as droplets forming fog or dew. Protection against water vapour under these conditions must consist of either a heated or a hermetically sealed (airtight) enclosure.

On a longer time scale, water can cause *corrosion* which can initiate early failure of an electronic system. Corroded connectors, for example, may not provide a sufficiently low-impedance ground or shielding connection to be effective against electromagnetic interference. Materials must therefore be selected carefully to suit the application environment. Anodized aluminium, for example, which is a cheap, attractive and popular material for benign indoor environments, soon oxidizes and loses its surface conductivity so that grounding connections become high-impedance joints, and are therefore ineffective.

Foreign Bodies

BS5490, mentioned in the previous section, also defines seven categories of protection against foreign bodies ranging in size from greater than 50 mm in diameter to dust. Each category in Table 9.3 is indicated by a numeral and can be combined with a numeral from Table 9.2 to form a two-digit code, prefixed with the letters IP. Thus IP45 means that an equipment enclosure so designated is protected against foreign bodies to category 4 (objects greater than 1 mm in diameter) and water to category 5 (water jets). Large foreign bodies are an obvious mechanical hazard inside an equipment enclosure, while smaller objects may cause damage if they enter switches or connectors, or if they are conductive. Similarly, dust can cause damage to exposed switch contacts and may also initiate current leakage or flashover. Measures to deal with these problems can only be empirical.

The classification scheme of BS5490 also indicates the degree to which internal parts are *accessible*. Fingers, for example, cannot penetrate an enclosure classified as IP3x or higher. Accessibility is discussed in Chapter 10 in connection with safety.

Mechanical Stress

Many electronic systems have to withstand mechanical stress in the form of impacts and shocks, vibration and acceleration. Some of these are amenable to analysis while others can be dealt with semi-empirically. Acceleration, for example, in an aircraft or missile will be known from the performance parameters of the aircraft or missile and the forces set up in a system by acceleration can be calculated from elementary mechanics. The effect on a component or system can then be calculated, or tested in a centrifuge.

Vibration, or periodic acceleration, however, is a much more difficult problem because vibrational energy may be coupled efficiently into components at certain frequencies by resonance, even to the extent of causing breakage by the application of excessive force to joints and fastenings. Some analysis may be possible, but testing is usually necessary to establish where breakages are likely to occur. This is likely to be a task for a vibration specialist with a background in mechanical engineering and shows that the design and development of high-performance electronic products or systems is a multi-disciplinary activity.

Quite often, of course, electronic design is not the major part of a project and electronics engineers are themselves working as specialists within a team.

Table 9.3   Degree of protection indicated by the first characteristic numeral of the IP classification: protection against foreign bodies and dust.

Second characteristic numeral	Degree of protection	
	Short description	Definition
0	Non-protected	No special protection
1	Protected against solid objects greater than 50 mm	A large surface of the body, such as a hand (but no protection against deliberate access). Solid objects exceeding 50 mm in diameter
2	Protected against solid objects greater than 12 mm	Fingers or similar objects not exceeding 80 mm in length. Solid objects exceeding 12 mm in diameter
3	Protected against solid objects greater than 2.5 mm	Tools, wires, etc., of diameter or thickness greater than 2.5 mm. Solid objects exceeding 2.5 mm in diameter
4	Protected against solid objects greater than 1.0 mm	Wires or strips of thickness greater than 1.0 mm. Solid objects exceeding 1.0 mm in diameter
5	Dust-protected	Ingress of dust is not totally prevented but dust does not enter in sufficient quantity to interfere with satisfactory operation of the equipment
6	Dust-tight	No ingress of dust

Extracted from BS 5490:1977 with the permission of the British Standards Institution

Humans

Before leaving the subject of environmental stress, it is worth pointing out that people are present in many application environments and that they can be as much of a hazard to a system (and themselves) as any of the factors so far discussed. The design of enclosures, controls, connectors and handles should take into account the likely level of skill and sympathy of the user, and they should be sufficiently robust to withstand the use and abuse which they are likely to receive. If an enclosure is big enough to sit or stand on, for example, then perhaps it should be designed to withstand a weight of at least 80 kg. A hand-held product such as a radio transceiver is highly likely to be dropped from a height of about 1.5 m and ought to be designed to withstand such a drop without major damage.

Drop tests are widely used for equipment likely to be dropped. If a hard unyielding surface such as steel plate or concrete is used, a drop from even 1 m can be a severe test.

Type Testing

This book has attempted to show that the design of successful electronic products and systems, while based on a large body of underlying theory, is still heavily dependent on the skill, intuition and foresight of the design team. Many factors are not amenable to precise analysis and some are not amenable to analysis at all. There is thus a need for some form of testing of a design before production. *Type testing* is carried out on a prototype or several prototypes with the aim of establishing whether the design meets its performance specification, and to discover any weakness or design errors so that corrective design, or development, can be undertaken before the system goes into production. On a complex design, type testing can be a very expensive and time-consuming job, especially when design changes are made as a result of testing and some tests have to be repeated to verify the behaviour of the modified design.

Environmental Testing

To test a system over its full range of operating temperatures, atmospheric pressures and humidities, environmental test chambers of varying sizes and degrees of sophistication are used. The smallest may be no more than a fan-assisted oven/refrigerator, while the largest may be big enough to contain a vehicle. The number of combinations of environmental parameters that can be applied to a system under test is likely to be limited by the expense and time involved so that some estimate must be made of the worst-case combinations of static conditions. Once one starts to consider dynamic conditions the number of possible tests becomes even greater, and some pragmatic judgement has to be exercised to determine what the product is likely to encounter in its operating environment. A few hours at 18°C and high relative humidity, followed by a fall in temperature to below the dew point, for example, would simulate conditions out of doors in a temperate zone at nightfall, and would test the product's ability to withstand condensation.

If the operating environment is accessible, *field trials* can be used to test a product using the real operating environment rather than a laboratory-simulated environment. A simulated environment provides more control over the testing and gives clearer results but can also be very expensive.

Consider, for example, simulating the operating environment of a geostationary communications satellite. In this case the real operating environment cannot be used for testing and the expense is unavoidable.

Electromagnetic Testing

Electromagnetic susceptibility and emissions tests are now becoming a common requirement for all products, even those intended for domestic and office applications. The severity of the tests applied can vary considerably, however. A missile system to be installed on a warship, for example, would be tested rigorously over a wide range of frequencies up to and including radar frequencies and at considerable field strengths, because of the proximity of the system to powerful antennae on board the ship. A cellular radio-telephone for use in a road vehicle, on the other hand, would be given a much less rigorous test, concentrating on the frequency range likely to be encountered on motorways and in towns. Powerful sources in the VHF band, for example, such as police vehicles, could be encountered, so that susceptibility to frequencies of around 100 MHz would certainly be tested.

Electromagnetic testing is carried out in EM test chambers, fully or partially lined with cones of radio-absorbing material (RAM) to absorb EM waves incident on the walls and prevent the creation of standing waves which would distort the intended field distribution in the chamber.

Exercise 9.3 Explain why the radio-absorbing material lining an EM test chamber is shaped into cones.

A large test chamber is illustrated in Fig. 9.1. Small chambers may be no larger than a small bedroom. The test chamber provides a controlled EM environment, excluding ambient EM fields and waves so that emissions from the system under test can be identified unambiguously, and containing powerful EM fields generated within the chamber for susceptibility testing. A considerable quantity of expensive test equipment is needed to detect, measure and generate EM fields, including antennae, spectrum analysers, swept frequency generators, r.f. power amplifiers and broadband receivers. Power and signals are fed in and out of a chamber through waveguide ports or through deep tubular ports acting as waveguides below cutoff. Doors must be sealed against EM fields by sprung metal fingers to ensure low-impedance grounding of the door to the chamber walls. Similar sealing is needed around the door of a microwave oven to contain hazardous microwave radiation within the oven.

Electronic Production Testing

Once a design is in production, other types of test are required to verify that each unit manufactured works correctly and is free from manufacturing faults which could cause early failure. There are two main methods of testing electronic products for correct operation and a variety of tests which expose the products to stress with the aim of detecting weak units.

In-circuit Testing

A common test technique for newly assembled PCBs aims to establish that every component on the PCB is correctly installed, of the right type or value, and functioning correctly. At first sight this might seem a superfluous test, but in a production environment it is very difficult to avoid assembling an occasional IC back-to-front, or a polarized component the wrong way round, or even mixing up components so that resistors of the wrong value, for example, are assembled into a board.

An in-circuit test machine carries out an electrical test on each component to verify its behaviour, value and orientation. The effect of neighbouring components is avoided by *guarding* or *back-driving*. Figure 9.2 shows a simple example of the technique. An in-circuit tester is connected to the circuit shown at A and B with the aim of checking the value of R_3. To do this a voltage of, say, 5 V is applied at A with respect to B. The tester measures the current flowing, I, to obtain the value of R_3 from $5/I$. In the circuit shown, however, the transistor's

Fig. 9.1 A large test chamber for electromagnetic testing, partially lined with cones of radio-absorbing material (RAM). The antennae shown are being set up for a calibration measurement. (Courtesy of Electromagnetic Engineering Department, British Aerospace, Filton)

base–emitter junction will be forward biased by the voltage at A and some current will flow through the base–emitter diode and R_4 to B. To overcome this, the base–emitter diode can be reverse biased by applying a voltage at D of say 6 V. Finally, current can flow from A through R_1 to the positive supply rail and through other circuits to 0 V and hence back to B. This can be prevented by applying the same voltage at C as is applied at A (from a separate source, not by connecting A and C together!) so that no potential difference exists across R_1 and therefore no

Care has to be taken as the voltage applied at D could be more than the normal voltage present when the circuit is operating and this could *overstress* R_4 by dissipating a higher power than the resistor is rated to handle.

153

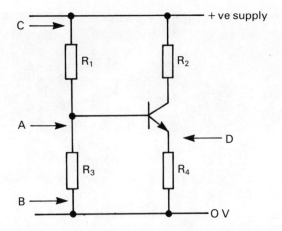

Fig. 9.2   An example circuit illustrating the principle of in-circuit testing.

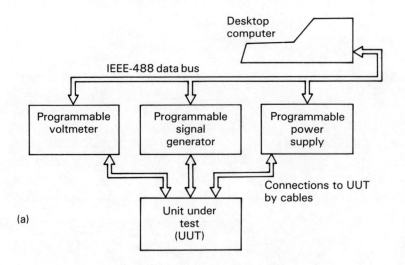

(a)

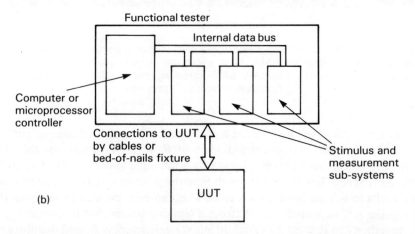

(b)

Fig. 9.3   Functional ATE: (a) programmable instruments controlled by a desktop computer *via* an IEEE-488 interface bus; (b) a functional tester.

current flows through R_1 (or R_2 via the base–collector junction of the transistor). This is the same technique of *guarding* which was discussed in Chapter 7 in the context of reducing leakage currents in a sample-and-hold circuit. One problem with in-circuit testing is the possible overstressing of neighbouring components, such as R_4 in the example above. This is normally dealt with by applying the test voltages for a short time only.

Functional Testing

An in-circuit test does not check that a PCB will actually work because there are a variety of faults such as bad joints, solder bridges or splashes between tracks and faulty plated-through holes which are not tested by an in-circuit test. A functional test checks the behaviour of the board with power applied and with test signals fed in. This type of test can be done manually using laboratory test gear such as oscilloscopes, logic analysers and signal generators or automatically using *automatic test equipment* or *ATE*. ATE can consist of conventional laboratory test equipment fitted with *interfaces* such as the IEEE-488 bus and controlled by a desktop computer, or for higher-volume testing a *functional tester* which is a computer or microprocessor-based system containing stimulus and measurement sub-systems equivalent in function to conventional test equipment. For testing small batches of boards the *unit-under-test* or *UUT* can be connected to the test equipment using *ad hoc* cables and connectors connected by hand. When larger quantities of boards are to be tested a more cost-effective and faster method of connecting to the UUT is a *bed-of-nails fixture*, as shown in Fig. 9.4. The UUT is held on the fixture by a partial vacuum, and electrical connections to the UUT are made by spring-loaded contacts or *nails* touching directly onto the PCB tracks. The ends of the nails are sharp so that they cut through any surface oxide on the solder to make a low-resistance connection to the board. Electrical connections from the nails are brought out to an airtight connector and thence through a cable to the ATE system. Bed-of-nails fixtures have to be custom-made and wired for each type of UUT, but the cost can soon be recovered by the saving in time obtained by their use.

The IEEE-488 bus is an 8-bit parallel interface capable of working at data transfer rates of 10^6 bytes (1 byte = 8 bits) per second over distances of up to 15 m.

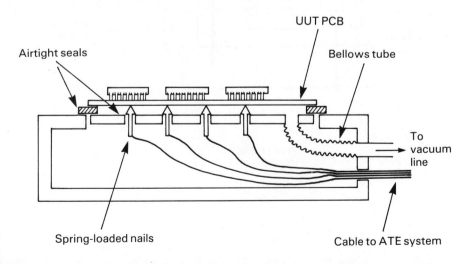

Fig. 9.4   Main principles of a bed-of-nails test fixture.

Typically, a board can be tested in 1–2 minutes using functional ATE and bed-of-nails fixtures.

Wilkins discusses *design for test* extensively in the field of digital circuits and gives references to recent literature on the subject.

Design for Testability

The use of ATE for product testing is cheaper and easier if the designer considers *testability* at the design stage. There is an increasing awareness in industry that electronics designers must consider how their design will be tested right at the start of the design process, rather than leaving test problems to be resolved when the design is almost complete.

Design for testability can consist of very simple and mundane measures such as ensuring that there are suitable places on the underside of a PCB where contact can be made by a bed-of-nails fixture, to more elaborate circuit design techniques to make testing easier. Figure 9.5 shows an example of circuit design for testability. Synchronous digital circuits are controlled by a *clock* or oscillator. Testing of a synchronous circuit normally requires that the circuit be clocked by the ATE system. The figure shows how, with the addition of a 2:1 data selector and two resistors, the testability of the circuit can be improved. In normal operation the clock select input and the tester clock input are left unconnected and the system clock input to the data selector is output to the rest of the system. When connected to the ATE system, the ATE clock is connected to the tester clock input and the ATE controls the clock select input so that either the ATE clock or the system clock can be used to clock the system. The *test point* is a terminal or output where the ATE system can monitor the system clock. Test points can be provided on a system to be tested manually to allow easy connection of test equipment. They may be connectors or special terminals soldered directly into a PCB.

Digital circuit testing is a non-trivial problem and much research effort has been

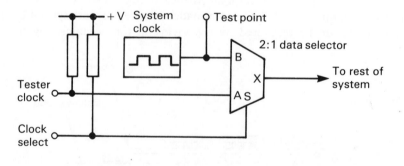

Truth table for
data selector

S	X
L	A
H	B

Fig. 9.5 An example of design for test: the clock circuit of a synchronous digital system.

devoted to devising circuit structures which are capable of being tested quickly. Even so some digital circuits cannot be tested thoroughly at all.

Stress Screening

Once a PCB or system has passed a functional test and has been found to work correctly, many manufacturers apply a final series of tests to expose *latent defects* in the unit which are not readily apparent, but which will cause early failure. The reason for carrying out these tests was explained in Chapter 8 in the context of the 'bathtub' failure rate curve. Latent defects can be detected by operating a product, possibly at an elevated temperature, for a period of days or weeks. This is known as *burn-in* or *soak testing*. A more effective method, however, is to subject the units to be tested to a series of carefully chosen *stress tests* or *stress screens*. These tests need not be elaborate or time consuming, but their effectiveness must be monitored by recording test data for each unit tested and collecting reliability data on the units which fail in service. Table 9.4 presents a summary of some possible stress tests used in electronic product testing. Some of the tests can be applied very easily. Powering the UUT up and then down (switching it on and then off), for example, stresses circuits because of *inrush currents* and *voltage surges* which are not present in normal operation and can cause weak components to fail.

Wilkins discusses techniques such as SCAN design, which enable synchronous digital circuits to be tested easily. Research work in this field is still continuing.

Stress screening is expensive and is normally applied only to high-value products or to systems where high reliability is essential.

Table 9.4 A selection of stress tests for electronic products

Test	Procedure	Detects
Power up/down	Connect the unit to a power source and switch on and off a few times, then test for continued function	Weak fuses and semiconductors
Thermal cycle	Run at elevated temperature then at low or room temperature and repeat	Weak and strained joints
Swept frequency vibration	Mount unit on vibration table and subject to vibration at a single frequency swept through a suitable range repeatedly	Loose fastenings, poorly mounted components
Random vibration	As previous test but using random vibration	More effective than previous test because of excitation of resonances

Summary

A selection of environmental factors influencing the performance and reliability of electronic systems has been discussed in this chapter. Temperature affects the parameters of many electronic components significantly, and design of precision circuits to operate over a wide temperature range is not easy. Component ageing is accelerated by increased temperature, so that the operating life of systems working at high temperature is reduced. The failure rate of many electronic components obeys the Arrhenius equation, which is a relationship between absolute temperature and the rate of a chemical reaction. One useful effect of the increased failure rate with increasing temperature is the practicality of performing accelerated ageing tests on components by operating them at high temperatures. The effect of water and dust on electronic equipment cannot be dealt with analytically. An empirical classification scheme for the degree of protection provided by an equipment enclosure against water and dust has been described. Vibration, acceleration and people have been briefly discussed as environmental hazards to electronic systems.

The need for environmental and electromagnetic testing, and the difficulty of testing exhaustively for all possible conditions, has been outlined. The chapter has concluded by introducing the three main types of test applied to production units: in-circuit testing to verify that components have been correctly assembled into a circuit; functional testing to verify that a system operates as intended; and stress testing or screening to detect units with latent manufacturing or component defects.

Problems

9.1 A capacitor has a failure rate of 8×10^{-6} hour^{-1} at 20°C and 2×10^{-5} hour^{-1} at 40°C. If the Arrhenius equation holds for this component what would be the failure rate at (a) 10°C and (b) 60°C?

9.2 An integrated circuit has a failure rate of 6×10^{-9} hour^{-1} at 20°C. (a) Express the equivalent MTTF in years. If the Arrhenius equation applies and the failure process has an activation energy of 0.25 eV what is the MTTF in years at (b) 0°C and (c) 70°C?

Safety

Objectives

☐ To emphasize the safety responsibilities of design engineers.
☐ To stress the importance of protection against electric shock and to outline the international safety classifications for electrical and electronic equipment.
☐ To discuss briefly some other risks posed by electronic equipment.
☐ To describe some of the design implications of safety standards and legislation.
☐ To study briefly the special safety measures required in two applications areas: medical equipment and equiment for use in potentially explosive or flammable atmospheres.

Modern electronic equipment has to be designed for safe operation both in normal use and under fault conditions and also during transit, installation and maintenance. Safety requirements and recommendations for electronic equipment are contained in a number of technical standards and codes of practice issued by the International Electrotechnical Commission (IEC), the British Standards Institution (BSI) and the European Committee for Electrotechnical Standardization (CENELEC).

Unsafe equipment may cause injury or death to the user, resulting in legal action against the equipment manufacturer. In recent years UK Acts of Parliament have been enacted which place legal responsibility for safety on equipment designers and installers among others. An electronics engineer who fails to take reasonable care over the safety of his design is committing a criminal offence. The most wide-ranging of these recent Acts is the Health and Safety at Work etc. Act 1974, which deals with the health and safety of those at work and includes within its scope any machinery, equipment or appliance designed for use in a place of work. Specific responsibilities are placed upon designers (and others) to ensure that equipment is safe when used properly, that adequate information is available about the equipment and any measures necessary to ensure its safe use, and to carry out research to discover, eliminate or minimize risks posed by the equipment. Electronic equipment designed solely for domestic use and not in a place of work is subject to the provisions of the Electrical Equipment (Safety) Regulations 1975 and the Electrical Equipment (Safety) (Amendments) Regulations 1976, which empower the Secretary of State for Trade and Industry to prohibit the sale or offer for sale of unsafe goods. Safety requirements for mains-operated electronic equipment and electrical appliances for domestic use are defined in British Standards 415 and 3456, and guidance on construction of electrical equipment to minimize the risk of electric shock is contained in a Memorandum published as BS2754. In Europe, mains-operated domestic equipment is approved by a CENELEC member organization. In the UK, approvals are granted by the British Electrotechnical Approvals Board (BEAB) after tests on a sample of the

CENELEC stands for Comité Européen de Normalisation Electrotechnique. The organization is based in Brussels and comprises the national electrotechnical committees of the EEC and Austria, Finland, Norway, Sweden and Switzerland.

For a useful general introduction to product safety and liability see Abbott.

This list of responsibilities is not exhaustive.

Information on the effects of electric current on the human body, including the threshold of perception, is given in IEC Publication 479: 1984.

equipment. The BEAB approval symbol which is marked on approved equipment is shown in the margin. Equipment tested by another CENELEC approvals authority carries the additional wording *via CCA* to indicate that BEAB itself did not carry out the tests. (CCA stands for CENELEC Certification Agreement.)

Electric Shock

The most common and dangerous risk posed by electronic equipment is electric shock. Design measures to minimize the risk of electric shock are therefore of fundamental importance. At mains frequencies, alternating currents of only 0.5 mA can cause a reaction in healthy people, while 50 mA can be lethal if sustained for more than one second. Figure 10.1 shows a range of currents and time durations and the regions in which physiological effects occur. Zones 3 and 4 in the figure represent dangerous combinations of current and duration. Even non-lethal shocks can cause injury because of involuntary action such as sudden recoil from the source of shock. Protection against electric shock can consist of measures to limit currents to safe levels, irrespective of voltage. Where current is limited to a safe level, contact with live high-voltage conductors can be quite safe. For

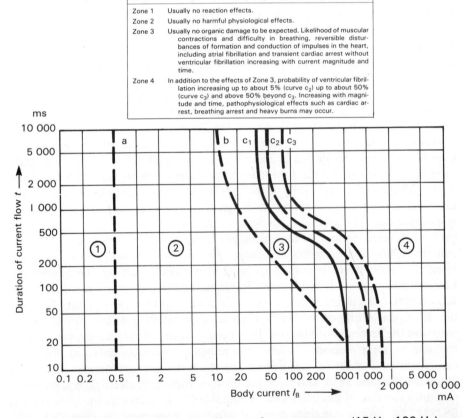

Zones	Physiological effects
Zone 1	Usually no reaction effects.
Zone 2	Usually no harmful physiological effects.
Zone 3	Usually no organic damage to be expected. Likelihood of muscular contractions and difficulty in breathing, reversible disturbances of formation and conduction of impulses in the heart, including atrial fibrillation and transient cardiac arrest without ventricular fibrillation increasing with current magnitude and time.
Zone 4	In addition to the effects of Zone 3, probability of ventricular fibrillation increasing up to about 5% (curve c_2) up to about 50% (curve c_3) and above 50% beyond c_3. Increasing with magnitude and time, pathophysiological effects such as cardiac arrest, breathing arrest and heavy burns may occur.

Fig. 10.1 Time–current zones of effects of a.c. currents (15 Hz–100 Hz) on persons. Reproduced from IEC Publication 479–1:1984 by permission of the International Electrotechnical Commission which retains the copyright.

Table 10.1 Voltage ranges generally applicable to electronic equipment

Designation	Between conductors		With respect to earth	
	V a.c. r.m.s.	V d.c.	V a.c. r.m.s.	V d.c.
LV Low voltage	< 1000	< 1500	< 600	< 900
ELV Extra-low voltage	< 50	< 120	< 50	< 120
SELV Safety extra-low voltage[1]	< 50	< 120	< 50	< 120

[1] An SELV supply must be isolated from the supply mains and earth by a safety isolating transformer, or be derived from a source, such as a battery, which is independent of a higher-voltage supply.

example, power-supply designs for CRTs producing up to 30 kV are safe if the design inherently limits current, typically to 200–300 μA.

When discussing safety measures against electric shock there is a need for precise definitions of voltage ranges. Table 10.1 lists the three voltage ranges defined by the IEC which are generally applicable to electronic systems. Note carefully that the term *low* does not mean safe: these voltages are only low relative to higher voltage ranges. Mains supply voltages in common use (110 V, 240 V) are classed as LV, but can be lethal. Extra-low voltages less than 50 V a.c. r.m.s. at mains frequencies, or less than 120 V d.c. are normally considered safe under dry working conditions. In wet or highly conducting locations these voltages are not necessarily safe and other precautions may be needed to prevent danger.

To some extent, the impedance of the human body limits current at low voltages. Current flow from one arm to the other across the chest, however, is especially hazardous because of the danger of electrical interference with the normal action of the heart. One should also be aware that in some applications there may be a risk to livestock as well as humans. Some farm animals are much more easily electrocuted than humans because of their low body resistance.

Protective Measures Against Electric Shock

There are four possible measures against electric shock which can be applied in the design of electronic equipment. First and most important is inaccessibility of *live parts*. A live part is defined as any conductor or conductive part intended to be energized in normal use, including a neutral conductor. The need to prevent access to live parts has many design implications, some of which are discussed later in this chapter.

The second protective measure against electric shock is *earthing* and automatic disconnection of supply in the event of a fault. The concept of an earth or ground is so often taken for granted in electronics that it is worth discussing the idea briefly from a safety viewpoint. When we state or measure a voltage or potential we refer to an arbitrary reference potential which we designate as 0 V. Quite often the 0 V reference is the protective conductor or 'earth' in the mains supply which is connected to the ground beneath us by a buried rod or plate. To be of any use for safety purposes the earth connection must be of low impedance, and provided this is so the potential of the earth connection will be effectively unaffected by currents

Information on protection against electric shock is contained in the IEE Wiring Regulations published as *Regulations for electrical installations*, 15th edition, 1981. Although the Wiring Regulations cover electrical *installations* rather than electronic systems many protective measures against electric shock discussed in the regulations are applicable to electronics. Interpretation of the regulations is given by Jenkins, B.D., *Commentary on the 15th edition of the IEE Wiring Regulations* published by Peter Peregrinus Ltd on behalf of the IEE, 1981.

flowing to or from earth because of the large mass of the ground. Any part of an apparatus (such as a metal case) which is securely connected electrically to earth and protected by a suitable automatic disconnection device, such as a fuse or circuit breaker, cannot become live at a dangerous voltage for a significant time. This principle is very widely used.

The third protective measure against electric shock, applicable in some cases, is limitation of voltage to SELV as defined in Table 10.1. An SELV supply is limited to less than 50 V a.c. r.m.s. and must be isolated from the supply mains and earth by a safety isolating transformer or other means. This is safe because these voltages are normally insufficient to cause a hazardous current to flow through the impedance of the human body.

A final protective measure against shock is limitation of current to a safe value, irrespective of voltage. This principle is used in flash testers for testing the insulation of electrical appliances at up to 4 kV. The tester is designed so that current is inherently limited to a safe value.

One final hazard should be pointed out under the heading of electric shock. Reservoir capacitors in electronic equipment can store significant electrical energy which can be hazardous if discharged through the body. Large-value capacitors must therefore be fitted with *bleed* resistors to dissipate the charge when the equipment is switched off and prevent injury to maintenance personnel.

Safety Classes for Electrical and Electronic Equipment

IEC publication 536 defines four safety classes for equipment with regard to protection against electric shock. These classifications do not indicate quality of protection but rather indicate how protection is achieved. Safety class 0 equipment has no protective measures against electric shock, and is therefore not safe unless used in a non-hazardous environment.

Safety Class I apparatus is designed to be connected to earth. All accessible conductive parts such as a metal panel, case or cabinet must be *bonded* electrically to a protective earth terminal. If the equipment is designed to be connected to a mains supply by a flexible cable, the cable must have a protective earthing conductor and be fitted with a plug with an earthing contact. Either the cable must be non-detachable or the apparatus must have a mains inlet connector with an earthing contact. Live parts must of course be inaccessible and there must be at least *functional insulation* throughout the apparatus. Functional insulation is defined as that necessary for proper functioning and basic protection against shock. Safety relies on protective earthing and automatic disconnection of supply to guard against failure of the functional insulation. Safety Class I equipment normally carries the legend: WARNING: THIS EQUIPMENT MUST BE EARTHED.

Safety Class II equipment has no provision for protective earthing and is normally *double-insulated* or equipped with *reinforced insulation*. Double insulation consists of functional insulation plus independent insulation to ensure protection against shock if the functional insulation fails. Reinforced insulation is a single layer of insulation which provides equivalent protection to double insulation. If the outer enclosure of the apparatus is durable and of insulating material it may be regarded as the second layer of insulation. If the outer enclosure is of metal, double insulation must be used throughout internally. Safety of Class II equipment does not depend on installation conditions such as correct wiring of

The buried rod or plate is known as an *earth electrode*. In the UK, in towns, this is normally located at the electricity board substation. Each customer supply cable includes an earth connection wired to the earth electrode.

Very heavy currents can alter the ground potential. For example if an earth fault occurs on a high-voltage transmission pylon, heavy currents flowing into the ground can raise the ground potential at the base of the pylon. It is possible for a cow or horse to be electrocuted if the animal stands radially close to the pylon.

The use of equipment of Class 0 is not allowed in the UK.

Bonded means connected together electrically to ensure a common potential. The wire or metal braid used must be of adequate cross-section to carry any fault current.

162

an earth conductor to a mains supply. Much household equipment is therefore designed as Class II. Examples are hairdryers; power drills; food mixers and table lamps. Double-insulated safety Class II equipment is marked with the double-square symbol shown in the margin.

Safety Class III equipment is designed for connection to an SELV supply and does not generate voltages higher than SELV. Protection against shock therefore relies on limitation of voltage. Handlamps for use in ship's bilges (a hazardous wet location) operate from a 25 V SELV supply.

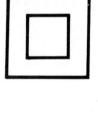

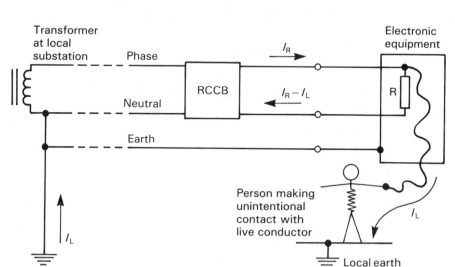

Fig. 10.2 Earth leakage.

Residual-Current Circuit Breakers

For added protection against shock, especially in workshops and electronics development laboratories, special mains circuit breakers are increasingly being used. Residual-current circuit breakers (RCCBs) detect small differences of the order of 10–30 mA in the currents flowing in the *phase* and *neutral* conductors of a mains supply due to leakage of current to earth (possibly through a human body in contact with a live conductor). The principle of operation of a common type of RCCB is shown in Fig. 10.3. The phase and neutral conductors connecting a load to a supply are wound in opposite senses on a toroidal core and a secondary detector winding is also wound on the core. The phase and neutral conductors form a primary winding, but the load currents in the two conductors are equal and opposite and generate no net magnetic flux in the core. Any difference current, however, generates an e.m.f. in the secondary winding which can be used to operate a trip coil to open the circuit breaker contacts.

RCCBs are also known as *earth leakage circuit breakers* or ELCBs. They are common examples of the wider class of *residual current devices* or RCDs. The provision of these devices in a workplace may be an example of an employer's responsibility under the Health and Safety at Work etc. Act 1974.

In Some RCCBs the phase and neutral conductors are simply threaded through the centre of the toroid, each forming a single-turn winding.

RCCBs must be tested regularly and are fitted with a test button for this purpose. Where RCCBs are installed in a workplace it is a good idea to turn off the mains supply at the end of each day by using the RCCB test button.

It is important to realize that RCCBs give no protection against contact with phase and neutral conductors simultaneously. Also the RCCB test button does not

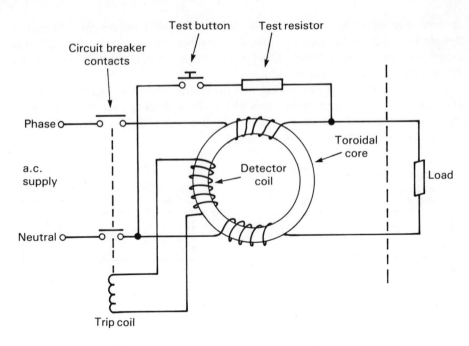

Fig. 10.3   A typical residual current circuit breaker (RCCB).

verify that the RCCB is operating within its performance specification: it merely checks the electromechanical function of the device.

Other Safety Hazards

Apart from the risk of death due to electric shock, electronic apparatus can be hazardous in other ways, a few of which are listed in Table 10.2. It should also be

Table 10.2   **A selection of safety hazards in electronic equipment**

Hazard	Source	Main risk
Ionizing radiation (mainly X-rays)	CRTs with accelerating voltages greater than 5 kV	Radiation exposure
Microwaves	R.f. circuits	Burns
Laser radiation	Lasers, laser diodes	Damage to eyesight
Toxic gases/fumes	Damaged or overloaded components	Poisoning
Sound/ultrasound	Loudspeakers, ultrasonic transducers, switched-mode power supplies	Hearing damage or loss
Implosion/explosion	Vacuum tubes, CRTs, overloaded capacitors and batteries	Injuries caused by flying glass or fragments
Heat	Hot components, heatsinks	Burns, fire.

realized that quite simple mechanical defects such as sharp edges and inadequate handles could result in injury.

Ionizing radiation in the form of X-rays can be produced in any apparatus which accelerates electrons through voltages greater than 5 kV. Cathode-ray tubes (CRTs) used in oscilloscopes, televisions, video monitors and visual display units (VDUs) are the most likely devices to generate X-rays. Limits on ionizing radiation are set by the International Commission on Radiological Protection (ICRP).

Hot surfaces on the outside of electronic equipment are an obvious hazard due not only to the risk of burns but also because unexpectedly hot surfaces, handles or knobs might cause a person to jump away suddenly or drop the equipment. A heat-sink may be allowed to reach a higher surface temperature than the rest of an enclosure, whereas knobs, handles and switches which are intended to be touched are likely to be limited to lower temperatures than general outer surfaces.

Design for Safety

So far in this chapter, electronic product safety has been discussed in the abstract. This section discusses some of the design implications of safety legislation and standards.

Inaccessibility of Live Parts

The need to make live parts inaccessible has many ramifications and often significantly influences the mechanical design of electronic equipment. Any removable cover or panel which exposes live parts must be fastened so that it can only be opened or removed using a tool. Openings for ventilation and so on must be small enough that access by fingers is prevented. A British Standard test finger is defined for checking openings where there is doubt as to whether a finger could touch a live part. An even more stringent requirement is that suspended foreign bodies should not touch live parts if introduced into an opening. This requirement could necessitate the provision of some sort of baffle under a ventilation opening or the addition of an insulating cover over live parts.

External plug and socket connections carrying voltages greater than extra-low voltage must be arranged so that any exposed pins or terminals of the connectors are on the dead side of the connection when the connector is separated. When a connector provides a protective earth connection as well as live supply connections, the earth terminal in the connector mates before the live terminals and unmates after them.

In many electronic equipment designs, all of the circuitry apart from the primary side of the mains transformer is at extra-low voltage. When the equipment is opened for servicing no mains conductors should be exposed. Wiring to the transformer or main on/off switch should be fitted with insulating sleeving where it is joined to terminals. If there are mains voltages on a PCB it is good practice to confine the mains wiring to as small an area of the board as possible and to provide an insulating cover over that part of the board, including the underside, labelled to warn of the danger. The same principles apply where high voltages are generated within a system: the high-voltage parts of the circuit should be partitioned and protected by insulating covers so that servicing personnel are not exposed to danger when the apparatus is under test.

X-rays are generated when high-energy electrons strike a target material. The mechanism is discussed briefly by Kip.

Take a look for yourself at a selection of mains connectors such as the BS4491 connector and the BS1363 13 A plug and check this statement.

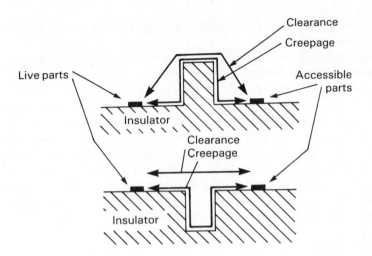

Fig. 10.4 Examples of creepage and clearance distances over an insulating barrier and a groove.

Creepages and Clearances

There are specified minimum distances between circuits connected to the mains supply and parts which are accessible. This is to allow for the possibility of a conductive path being established by dust and dirt where there would otherwise be no conduction. A *clearance* distance is the shortest distance measured in air between two conductive parts and a *creepage* distance is the shortest distance measured over the surface of insulation. As an example, BS 4743 gives a creepage distance of 3 mm for a 250 V a.c. r.m.s. voltage in a Safety Class I apparatus.

Mechanical Strength

All of the electrical safety features, such as inaccessibility of live parts, must be proof against reasonable mechanical abuse. A drop test as shown in Fig. 10.5 is one test which equipment might be required to withstand without breakage of

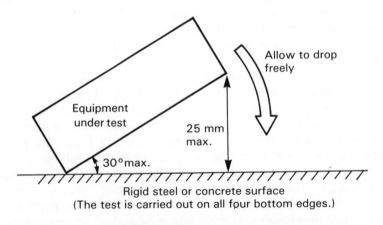

Fig. 10.5 Standard drop test.

insulation, cracking of the enclosure or loosening of covers. There may also be specified vibration and impact tests and requirements for mechanical integrity to be maintained while the equipment is exposed to heat.

Markings

To comply with safety standards, electronic apparatus must be marked with certain information including the nature of the electrical supply (a.c. or d.c.), the supply frequency if a.c., the rated voltage and the power consumption if more than 25 VA, the *off* position of the on/off switch if any, and the meanings of any indicator or warning lamps. The markings must be clear, suitably placed and indelible.

Protection Against Fire

When a fault occurs in an electronic system, heavy currents can flow, causing overheating of wiring and components. Faults can be due to component failure, mechanical damage, short circuits caused by foreign bodies falling into ventilation apertures, or deliberate abuse. Protection against fire caused by overloads is obtained by automatic operation of devices to interrupt the flow of excessive current.

The cheapest protective device is a fuse: a wire link designed to melt rapidly above a certain current and so break the circuit. Fuses are not foolproof and there is an incidence of bad practice by unskilled people replacing them with fuses of higher rating or even with improvised substitutes such as bent paper clips and nails. For electronic applications there are two main types of fuse. *Quick-acting* fuses consist of plain fuse wire or strip inside a glass or ceramic cartridge and are designed to rupture quickly once the *rated current* has been exceeded. *Time-lag* or *surge-proof* fuses are able to carry overload or surge currents of a few times their rated current for several hundred milliseconds. A common application for this type of fuse is in power supplies where there can be a large *inrush current* just after switching on as the reservoir capacitors charge up.

The *rupturing capacity* of a fuse is a very important performance parameter. Quick-acting fuses are available as *high-rupturing capacity* (HRC) types with a rated rupturing capacity of 1.5 kA, or as *low-rupturing capacity* (LRC) types with a rated rupturing capacity of 35 A or ten times the rated current, whichever is greater. Time-lag fuses are of the LRC type. If an LRC operates to disconnect a heavy current greater than its rupturing capacity the fuse wire may vaporize rather than melt and form a conductive path between the end terminals of the fuse link, either as an arc or along the walls of the fuse link, and the current may not be interrupted. HRC fuses incorporate additional design features such as a silica filling, to ensure that the arc resulting from operation by a heavy current is quenched and the current is safely interrupted.

Where the additional cost can be justified, perhaps on larger equipment, *circuit breakers* are to be preferred to fuses, as they cannot be tampered with by unskilled users. They have the additional advantages that they can be reset easily once the fault condition has been cleared and they can break more than one pole of the supply. The performance characteristics of fuses and circuit breakers are shown on time–current characteristics of the form illustrated by Fig. 10.6. Two curves on the

The replaceable part of a fuse assembly is known as a *fuse link*.

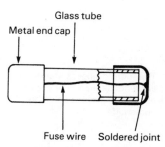

Typical construction of a quick-acting (LRC) fuse link.

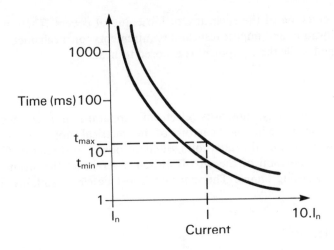

Fig. 10.6   A typical fuse time/current characteristic. I_n is the rated current.

characteristic show the maximum and minimum times for the device to operate as a function of fault current.

When a protective earth conductor is present it must *never* be disconnected by the action of any overload protection device.

Case Studies

Medical Electronic Equipment

The more stringent standards are a consequence of greater risk, not a greater value on human life.

Electronic equipment for use in medical and therapeutic applications is subject to much more stringent safety requirements than equipment for normal industrial use. Some indication of the stringency of the safety requirements can be given by stating that the first part of the relevant British Standard (BS 5724: Part 1: 1979) runs to over 200 A4 pages. This should be compared with safety standards for other application areas such as office equipment where less than 100 pages is sufficient to state the full safety requirements.

Apart from electric shock and other hazards which arise from all electronic equipment, medical apparatus can endanger life simply by failing to operate, either during normal use (as part of a life-support system, for example) or on demand for a device such as a defibrillator, which is used to restart a stopped heart by applying a controlled electric shock. Also, proper ergonomic design plays a significant part in the safety of medical equipment. Human error must be avoided as far as possible by design, especially because the most critical items of equipment are likely to be used in emergencies when staff are working under the greatest stress.

Ergonomics is the study of the interaction of people with their working environment.

Many of the hazards present in medical environments also occur in conventional industrial workplaces. There are however, special conditions in hospitals and ambulances which make the need for safety greater. Patients under anaesthesia or otherwise unconscious cannot react in the normal way to shocks and stimuli so that the presence of a fault or danger may not be apparent to staff. Patients may be connected electrically to monitoring equipment or even to more than one piece of equipment at a time. High-powered systems and sensitive instruments may be operated in close proximity so that electromagnetic compatibility is important. Operating theatres can be especially hazardous places because of the presence of moisture, humidity and sometimes flammable anaesthetic gases. Equipment designed to operate in an operating theatre will need to be proof against disinfection.

Electromagnetic compatibility was discussed in Chapter 7.

There are three levels of protection against hazard (sometimes referred to as *defence in depth*). The most important safety level is that of the equipment itself, which must have inherent safety features in its design. Examples are the use of double insulation and mains transformers and filters with earth leakage currents of less than 0.5 mA. Secondly, electrical installations in hospitals have safety features not found in normal electrical installations. These include safety isolating transformers with very low earth leakage currents, and special supplies designated medical safety extra-low voltage (MSELV) for operation of certain equipment. The final, and perhaps least reliable, safety measure is correct use of the equipment by trained personnel.

Reliability of medical equipment must be high and equipment must also be fail-safe: failure of one level of protection should not create a hazard and failures should be detected and indicated by an unmistakable fault indication. If this is not possible a periodic test and inspection procedure is necessary to detect faulty apparatus.

Equipment for use in Flammable Atmospheres

Electrical and electronic equipment for use in industries where flammable gases and liquids are present is subject to stringent safety requirements to reduce the risk of fire or explosion. High-risk industries include coal-mining, where methane gas can be released from coal seams and create an explosion hazard, and oil and gas handling and production where hydrocarbon fuels constitute a fire or explosion risk.

In the UK, equipment to be used in mines must be certified by the Mining Certification Service (HSE(M)). The British Approvals Service for Electrical Equipment in Flammable Atmospheres (BASEEFA) certifies equipment for use in other industries where potentially flammable or explosive atmospheres exist. Both organizations are part of the Health and Safety Executive (HSE) and are located at Buxton in Derbyshire.

As in other specialized technical fields, special terminology applies. Hazardous areas, in a petrochemical plant, for example, are classified into *Zones* according to the type and nature of flammable materials present and the degree of ventilation existing. An area classified as Zone 0 is a place where an explosive gas–air mixture is continuously present or present for long periods, and an area where an explosive gas–air mixture is likely to occur in normal operation is classified as Zone 1. An area where an explosive gas–air mixture is not likely to occur in normal operation, and if it does occur will be present for a short time only, is classified as Zone 2. Certified equipment carries a classification according to the type of protection against ignition, and the types of protection acceptable in each Zone are specified in Codes of Practice for the safe use of equipment in potentially explosive atmospheres.

The *Zone* terminology is defined in BS5345 which does not apply to equipment for use in mines.

Codes of Practice are often issued as British Standards. A Code of Practice sets out recommendations (or sometimes requirements) for the application or use of some methods, techniques or use of equipment.

The two most important types of protection against ignition (there are currently

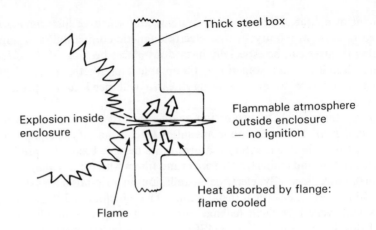

Fig. 10.7   Principle of a flameproof enclosure.

six others) are *flameproof enclosures* and *intrinsic safety*. All the methods of protection are designed to ensure a very low risk of the equipment igniting an explosive or flammable atmosphere. The most likely sources of ignition are electrical sparks caused by component faults or switch contacts, and overloaded components.

Flameproof enclosures achieve safety by containing and limiting an explosion to the inside of an enclosure. The electrical or electronic equipment is totally enclosed within a strong metal box designed to withstand an internal explosion without breaking or rupturing. The enclosure is not gas-tight: flammable gas can enter but an internal flame cannot ignite an external flammable atmosphere. All joints and openings in the enclosure have closely machined flanges of specified minimum depths so that flames are cooled by contact with the metal surfaces before reaching the outside of the enclosure and are then unable to ignite the external explosive atmosphere. Equipment to be certified as flameproof is subjected to rigorous tests, including actual explosions inside test specimens.

Electronic or light-current electrical equipment is often designed and certified

There is a minimum energy level below which a spark will not ignite a given flammable atmosphere.

for intrinsic safety. This approach ensures that all the electrical circuits in the equipment are incapable of causing ignition because insufficient energy is available. Limits on current voltage and power are closely specified and there are restrictions on circuit inductance and capacitance. Circuit faults have to be analysed to check that safety is not impaired and power-supply circuits have either to be intrinsically safe themselves or to be outside the hazardous area and have

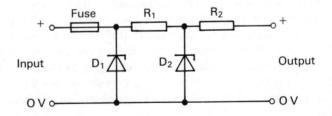

Fig. 10.8   Circuit diagram of a Zener diode safety barrier.

intrinsically safe outputs. A low-powered battery, for example, could be intrinsically safe or the equipment could be powered from a low-voltage mains supply. If power is drawn from outside the hazardous area there has to be some way of guaranteeing that the power and voltage limits for intrinsically safe equipment are not exceeded even in the event of a fault. This can be achieved by Zener diode *safety barriers* through which all operating power for the intrinsically safe equipment must be drawn. The safety barrier limits the voltage and current flowing into the hazardous area in the event of a fault and thus ensures that the conditions for intrinsic safety are maintained.

If a Zener diode safety barrier illustrated in Fig. 10.8 is a 10 V 47 Ω barrier (meaning that D_2 is a 10 V Zener diode and R_2 is a 47 Ω resistor), what is the maximum power deliverable to a load in the hazardous area? Ignore D_1 and R_1. (*Answer*: 0.53 W)

Exercise 10.1

Summary

Responsibility for the safety of an electronic system rests primarily with the equipment's designer. Neglect of product safety is a criminal offence in the UK.

The most important danger posed by electronic equipment is electric shock and protective measures against shock are most important. These include inaccessibility of live parts, protective earthing and automatic disconnection of supply, and limitation of voltage and/or current. RCCBs can provide additional protection against shock in laboratories and workshops.

Safety Class I equipment achieves protection against shock by having accessible conductive parts connected to a protective earth. Safety Class II equipment does not depend on protective earthing and is often double-insulated. Safety Class III equipment operates from an SELV supply.

Other safety hazards posed by electronic equipment include ionizing, microwave and laser radiation, sonic and ultrasonic pressure, heat and fire. Fuses and/or circuit breakers protect against fire caused by electrical overload.

Some of the design consequences of safety requirements have been discussed including the need to restrict openings and holes to prevent access to live parts and the need for mechanical integrity to ensure continued safety.

The special safety requirements of equipment for medical use and for use in flammable atmospheres have concluded the chapter.

Bibliography

Books

Abbot, H.: *Safe enough to sell?*; The Design Council (1980).

Anderson, J.C., Leaver, K.D., Alexander, J.M. and Rawlings, R.D.: *Materials Science*; Thomas Nelson (2nd edition, 1974).

Bradley, D.A.: *Power Electronics*; Van Nostrand Reinhold (1987).

Carter, R.G.: *Electromagnetism for electronic engineers*; Van Nostrand Reinhold (1986).

Cluley, J.C.: *Electronic equipment reliability*; Macmillan (2nd edition, 1981).

Compton, A.J.: *Basic electromagnetism and its applications*; Van Nostrand Reinhold (1986).

Keiser, B.: *Principles of electromagnetic compatibility*; Artech House (1979).

Kip, A.F.: *Fundamentals of electricity and magnetism*; McGraw-Hill (2nd edition, 1969).

Kraus, J.D. and Carver, K.R.: *Electromagnetics*; McGraw-Hill (2nd edition, 1973).

O'Connor, P.D.T.: *Practical reliability engineering*; Wiley (2nd edition, 1985).

Ritchie, G.J.: *Transistor circuit techniques*; Van Nostrand Reinhold (1983).

Scarlett, J.A.: *An introduction to printed circuit board technology*; Electro-chemical Publications (1984).

Senturia, S.D. and Wedlock, B.D.: *Electronic circuits and applications*; Wiley (1975).

Till, W.C. and Luxon, J.T.: *Integrated circuits: materials, devices and fabrication*; Prentice-Hall (1982).

Vincent, C.A. (ed.): *Modern batteries*; Edward Arnold (1984).

Wilkins, B.R.: *Testing digital circuits*; Van Nostrand Reinhold (1986).

Wong, H.Y.: *Heat transfer for engineers*; Longman (1977).

British Standards

The following list is a short selection of relevant British Standards. A full list and index of British Standards is published annually in the British Standards Institution Catalogue.

BS415: 1979, *Specification for Safety requirements for mains-operated electronic and related apparatus for household and similar general use.*

BS2754: 1976, *Memorandum: Construction of electrical equipment for protection against electric shock.*

BS3456: Part 1, 1974, *Specification for Safety of household electrical appliances.*

BS4743: 1979 (IEC348: 1978), *British Standard Specification for Safety requirements for electronic measuring apparatus.*

BS4778: 1979, *Glossary of terms used in Quality assurance (including reliability and maintainability terms).*

BS5490: 1977 (IEC 529: 1976), *British Standard Specification for Classification of degrees of protection provided by enclosures.*

BS5783: 1984, *British Standard Code of practice for handling of electrostatic sensitive devices.*

BS5850: 1981, *Specification for Safety of electrically energized office machines.*

BS6221: Part 21, 1984, *British Standard Printed Wiring Boards Part 21. Guide for the repair of printed wiring boards.*

BS6221: Part 3, 1984, *British Standard Printed Wiring Boards Part 3. Guide for the design and use of printed wiring boards.*

IEC Publications

IEC 536, *Classification of electrical and electronic equipment with regard to protection against electric shock.*

IEC 479: 1984, *Effects of current passing through the human body.*

Answers to Problems

2.1 3.4 mm^2
2.2 4.9 mm

3.1 6 m^2
3.2   250–500 million

4.1   About 200 mW
4.2   18 000 F!
4.3   High temperatures accelerate self-discharge.
4.4   (a) 2.5 mF
      (b) 5.1 A (no change)
4.5   (a) $C/8$
      (b) $3C/32$
4.6   (a) 70%
      (b) 10%
4.7   (a) 70%
      (b) 40%

5.1 $\pm 0.1\%$
5.2 $1/\pi$ pF or about 0.3 pF
5.3   13 mV r.m.s.

6.1   About 1°C W^{-1}
6.2   (a) 1.5 W
      (b) 83°C
6.3   21 m^3 hour^{-1}

7.1   Show that this model is equivalent to eqn (7.5) in terms of the voltage at the termination.
7.2   (a) About 1.5 GHz
      (b) -48, -38, -34, -31, -29 dB respectively for the 1st, 3rd, 5th, 7th and 9th harmonics (Use 20 $\log_{10}(|v_i|/V)$)
7.3   (a) $R_1 = 200\ \Omega$, $R_2 = 300\ \Omega$
      (b) 20×10^6 items per second
7.4   (a) 200 mA
      (b) 12 nF
      (c) No account has been taken of the time taken for charge to leave the power-supply reservoir capacitor.

8.1   (a) 99.76%
      (b) About 1100 hours
8.2   (a) 79 000 hours
      (b) 0.88
8.3   0.65
8.4   3.3 years

9.1   (a) 4.8×10^{-6}
      (b) 4.5×10^{-5}
9.2   (a) 19 000
      (b) 39 000
      (c) 4 500

Index